Mathematics Study Resources

Volume 22

Series Editors

Kolja Knauer, Departament de Matemàtiques i Informàtica, Universitat de Barcelona, Barcelona, Spain

Elijah Liflyand, Departament of Mathematics, Bar-Ilan University, Ramat-Gan, Israel

This series comprises direct translations of successful foreign language titles, especially from the German language.

Powered by advances in automated translation, these books draw on global teaching excellence to provide students and lecturers with diverse materials for teaching and study.

Jörg Neunhäuserer

Introduction to the Philosophy of Mathematics

 Springer

Jörg Neunhäuserer
Goslar, Germany

ISSN 2731-3824   ISSN 2731-3832   (electronic)
Mathematics Study Resources
ISBN 978-3-662-72178-0   ISBN 978-3-662-72179-7   (eBook)
https://doi.org/10.1007/978-3-662-72179-7

Acknowledgments At this point, I would like to thank my philosophy teachers at FU Berlin. I owe many insights to Peter Birie, Holm Tetens, Bernhard Thöle, Markus Otto, and Ursula Wolf, which cannot be found in any curriculum.

Of course, my thanks also go again to my parents and my wife, Katja Hedrich, who by now probably knows more about mathematics and its philosophy than she ever wanted to know.

My special thanks also go to Andreas Rüdinger and Springer Publishing, who made this book possible.

Competing Interests The author has no competing interests to declare that are relevant to the content of this manuscript.

<table>
<tr><td>Goslar, Germany</td><td>Jörg Neunhäuserer</td></tr>
<tr><td>2025</td><td></td></tr>
</table>

Contents

Introduction 1

If we understand philosophy as a rational reference to the world as a whole, then a philosophy of mathematics is a rational reference to mathematics as a whole. We expect such a reference to be understandable and meaningful in every case, ideally, it is also justified. A philosophy of mathematics appears understandable to us if the terms used are clearly explained or at least satisfactorily interpreted and if central statements are clearly and unambiguously formulated or can at least be formulated in this way. For a philosophy of mathematics to be meaningful, it must necessarily be free of contradictions. Contradictory statements give us no information and, according to classical logic, anything follows from them. The meaningfulness of an understandable and contradiction-free philosophy is measured by whether it is able to answer philosophical questions. With regard to mathematics, the ontological question first arises as to what kind of objects mathematics studies and in what sense these objects exist. In what sense do sets, relations, functions, numbers, spaces, geometric figures, etc. exist and what kind of objects are these? The second question that arises is the epistemological question of how we acquire mathematical knowledge and what kind of justification of mathematical statements should be accepted. Why are we allowed to count the statements of mathematical theories as our knowledge and how can these statements be justified? The third question that arises is the scientific-theoretical question of the relationship between mathematics and the other sciences. What is the connection between mathematics and physics, between mathematics and computer science, etc.? One might also ask ethical and aesthetic questions with regard to mathematics, but these have been of lesser importance in the history of the philosophy of mathematics. Once an understandable and meaningful philosophy of mathematics is available, the question of its justification arises. It is too much to expect a final justification of a philosophy of mathematics that answers ontological or epistemological questions. Nevertheless,

J. Neunhäuserer, *Introduction to the Philosophy of Mathematics*,
Mathematics Study Resources 22, https://doi.org/10.1007/978-3-662-72179-7_1

based on basic intuitions and scientific findings, one can argue for or against a position in the philosophy of mathematics. Such arguments are an essential part of the philosophy of mathematics, as they reveal logical relationships between philosophical statements.

In this book, we present the major positions in the philosophy of mathematics. We strive to make the To formulate the essence of a position in the simplest possible theses. We will explain these theses, discuss their interpretations, and specify variations. We always delve into the bibliographical background; the reader learns which philosophers and works a position in the philosophy of mathematics can be traced back to. If we find this particularly interesting or relevant, we also sketch the biographical or historical context of a position. This book does not offer a history of the philosophy of mathematics. We are more concerned with the content of philosophical positions and their systematic classification than with their historical genesis and impact. In some cases, a philosopher's philosophy of mathematics is inseparably linked to his overall philosophical design. In these cases, we offer a brief summary of relevant aspects of a philosophical system. Often, concepts and results of logic and basic mathematical disciplines such as set theory are essential for the philosophy of mathematics, in some cases the development of advanced mathematical theories is also of importance. It is a balancing act to convey all the crucial facts without overwhelming the reader with mathematics. We have tried to keep the indispensable mathematical explanations in this book understandable even for a non-mathematician, and can only hope that we have succeeded.

Our first goal in this book is the value-free presentation of key positions in the philosophy of mathematics, our second goal is the critique of these positions. Descriptive sections alternate in each chapter with sections that contain argumentation, evaluations and even some polemical remarks. We do not pretend to be impartial in these sections, our perspective is determined by our position in the philosophy of mathematics. We are proponents of a Platonic realism in the ontology of mathematics, a rationalism in the epistemology of mathematics, and a scientific demarcation of mathematics from other sciences. This means that we believe that the objects of mathematics exist independently of mental processes and beyond physical space-time, that we gain mathematical knowledge through immediate rational insight and logical deduction, and that mathematics is clearly distinguished from all other sciences by its methodical practice. This position is not original, rather conservative and controversial in the philosophy of mathematics. Perhaps our perspective is typical for a mathematician who deals with the philosophy of mathematics. In any case, our position seems to be more acceptable among mathematicians than among philosophers. We are not saddened if some readers join our position after reading this book. But we are just as pleased if our evaluations and arguments provoke readers and encourage contradiction. If this book makes a small contribution to a reader being able to formulate and justify his own position in the philosophy of mathematics, it fulfils its purpose. For the reader who is interested in getting to know other perspectives on the topics discussed in this book, we recommend reading the essays in the *Stanford Encyclopedia of Philosophy* and the *Oxford Handbook of Philosophy of Mathematics and Logic*.[1] Both sources were helpful and stimulating in the preparation of this book.

To the Second Edition
In this expanded version of the book, readers will find a new chapter on mathematics in German Idealism.

---

[1] See Zalta (2017) and Shapiro (2007).

Pythagoreanism

2

Contents

2.1 Pythagoras and the Pythagoreans

Pythagoras was already a legendary figure in antiquity; the sources are not entirely clear in terms of his biography.[1] Pythagoras was probably born in 570 BC on the Greek island of Samos in the eastern Aegean near the coast of Asia Minor. On the mainland not far from Samos was the ancient Greek city of Miletus, which is considered a cradle of science, since there the natural philosophers Thales (about 624–546 BC), Anaximander (about 610–547 BC) and Anaximenes (about 586–526 BC) worked. As a young man, Pythagoras is said to have travelled to Egypt and Babylon and got to know the high cultures there. Around 538 BC Pythagoras fled from the tyranny of Polykrates (about 570–522 BC) from Samos and settled in Greek-colonised southern Italy. In the city of Croton, he probably founded the religious-philosophical order of the Pythagoreans around 530 BC. Pythagoras was the master of this community and his word had unrestricted authority. The Pythagoreans were also politically active in southern Italy and came into conflict with other groups. Around 500 BC, Pythagoras had to flee from Croton and moved to Metapontum, which became the new centre of the Pythagorean community. There

---

[1] An important source is the history of philosophy of the ancient philosophy historian Diogenes Laertius from the third century AD, see Diogenes (2009). However, the presentation by Diogenes Laertius does not always agree with older sources, see Mansfeld (1986).

J. Neunhäuserer, *Introduction to the Philosophy of Mathematics*,
Mathematics Study Resources 22, https://doi.org/10.1007/978-3-662-72179-7_2

he probably died after 495 BC. Until the fourth century BC, the Pythagoreans had political influence in southern Italian cities, by the end of this century the movement seems to have died out. In the first century BC, there was a rediscovery of Pythagorean thought in the Roman Republic and a neo-Pythagorean current in philosophy, which remained effective until the second century AD.

Neither Pythagoras nor his immediate students left any writings. It is even reported that the Pythagoreans were encouraged to keep the teachings of their master secret. Therefore, it is not possible to determine unequivocally what Pythagoras taught. Even in antiquity, different representations of the Pythagorean doctrine circulated and historians still argue today about which doctrine we can attribute to Pythagoras. Some see Pythagoras as a religious leader, who was granted infallible access to divine knowledge. Others see him as a natural philosopher and mathematician, who sought a comprehensive understanding of the world.[2] In any case, the Pythagoreans were followers of a reincarnation doctrine, according to which the soul is immortal and is reborn in another living being after death.[3] This doctrine was not widespread in Greek culture at the time of Pythagoras, rather, the Homeric view that the souls of the dead continue to exist as shadows in the underworld was popular. In connection with the doctrine of reincarnation, religious rules and rituals of the Pythagoreans are handed down. For example, the Pythagoreans were encouraged to protect the lives of sentient beings and to eat a vegetarian diet. Legend has it that Pythagoras even possessed the ability to communicate with animals and remember past incarnations. These abilities are not typically part of the core competence of a philosopher and mathematician. Following the account of the Greek philosopher Aristotle (384–322 BC), Pythagoras appears to us in a different light.[4] Among other things, he is said to have developed a cosmogony and cosmology, that is, a doctrine of the origin and nature of the cosmos, based on natural numbers and their ratios. We will go into this in more detail in the next section. Furthermore, it can be assumed that Pythagoras actually knew the famous theorem that bears his name.[5] See Fig. 2.1 for this. Also, Pythagorean triples, i.e., natural numbers a, b, c with $a^2 + b^2 = c^2$, such as 3, 4, 5 or 5, 12, 13, are said to have been studied by Pythagoras. Whether Pythagoras knew a proof of Pythagoras' theorem and whether he discovered the theorem himself or learned it on his travels to the Orient cannot be said with certainty. Nevertheless, mathematics and its philosophy seem to us to be just as fundamental to the worldviews of the Pythagoreans as the doctrine of reincarnation.

---

[2] The dispute among historians has led to numerous publications, we refer here to the entry on Pythagoras in Zalta (2017) and the bibliography in this article.

[3] The crucial point here is fragment 7 of Xenophanes, who was a contemporary of Pythagoras, see Heitsch (2014).

[4] See Aristoteles (2003).

[5] This is disputed among historians. We rely here on Proclus' comments on Euclid's Elements, see Mansfeld (1986).

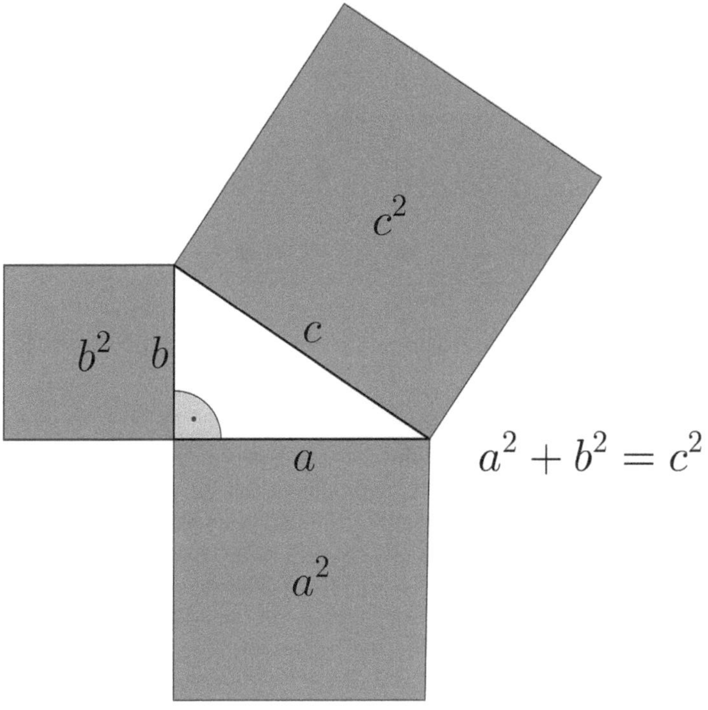

Fig. 2.1 Pythagoras' theorem

2.2   Ontological Pythagoreanism

A strong or ontological Pythagoreanism in philosophy is characterised by the following principle:

Mathematical objects are ontologically fundamental, the world consists of them.

This thesis obviously requires further explanation. Which mathematical objects are meant and what properties do they have? In what way do these objects constitute the world? We will first present the position of the Pythagoreans, which Aristotle describes in his Metaphysics.[6] Afterwards, we will attempt to provide a contemporary explanation of ontological Pythagoreanism, which may appeal to some readers.

The famous Pythagorean thesis *Everything is number* means that natural numbers and their ratios are the basic building blocks of the world. According to the Pythagorean cosmogony, in the beginning was the One, as the limited and limiting

---

[6] See Aristoteles (2003).

Fig. 2.2 The Tetraktys

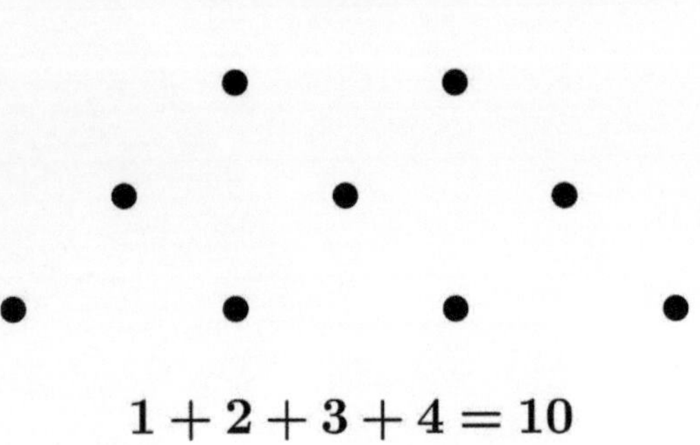

$$1 + 2 + 3 + 4 = 10$$

in the unlimited (*Aperion*). It is the origin of the world. The One has the ability to reproduce itself by division. It breathes in some Aperion, doubles itself and becomes Two. In the next step, Three is created, and so on. To understand this, we must consider that the Pythagoreans identify numbers with sets of points and imagine these spatially. One is ●, Two is twice One with some emptiness in between, so ● ●. Three is accordingly ●●● etc. Ten is identified with the Tetraktys and has special significance for the Pythagoreans, see Fig. 2.2. For the Pythagoreans, it is the *source and root of the ever-flowing nature* and represents the cosmic totality. With the Tetraktys, the harmonic ratios 2:1, 3:2 and 4:3 are created simultaneously, i.e., the octave, the fifth and the fourth. For the Pythagoreans, these are not only the basis of music, but also the basis of the cosmos. This is optimally ordered by the harmonies and the cosmology shows the same laws as in music. The Pythagoreans speak of the *harmony of the cosmic spheres* and legend has it that Pythagoras could hear this. Natural numbers are not only understood as quantities; rather, they are attributed quantitative properties and real powers. The number Two is the opposition, the difference and the dispute. The number Three represents the cycle of birth, life and death or beginning, middle and end. The even numbers represent the female and the odd numbers the male force. Five, as the sum of Two and Three, is the marriage or community of the female and male. It is probably fair to say that the Pythagoreans were inclined towards a number mysticism.

The cosmogony and cosmology of the Pythagoreans may be of certain poetic charm. However, if we base our understanding on the findings of contemporary cosmology, they are unacceptable. Some natural numbers and rational ratios are certainly not sufficient as a description of the universe and its origin. More interesting is the philosophy of mathematics of the Pythagoreans. For them, the objects of mathematics, in particular numbers, are not abstract but concrete. They are physically realised and effective. This view will reappear at several points in this book. In contemporary ontological naturalism, in concrete structuralism and also in certain forms of formalism we find similar principles. See Chap. 9, Sects. 11.3 and 12.4 for this. The crucial difference between the attitude of the Pythagoreans and these approaches in the philosophy of mathematics lies in the Pythagorean assumption that the objects of mathematics are ontologically fundamental and constitute the world as a whole. In all other philosophical conceptions, concrete mathematical objects, structures or formulas are found alongside other concrete objects of the world. If we are willing to make strong assumptions, an ontological Pythagoreanism can also be formulated in a contemporary way. First, we assume that the world can be completely described by mathematically formulated theories. There are scientists who are actually convinced that this could be possible. In particular, physicists are looking for a world formula or a theory of everything. Such a theory is supposed to unify the four known physical forces and describe both the properties of elementary particles, i.e., the world on a small scale, as well as the properties of the world on a large scale, i.e., the universe.[7] If such a theory of everything is available in the language of mathematics, we could try to identify all objects of the world with objects in our theory. Philosophically, this is referred to as an ontological reduction. Just as water is nothing but H_2O, elementary particles at a fundamental level could be nothing but probability distributions on a Hilbert space or other objects in our mathematical formalism. If the things to which we reduce everything were indeed mathematical, we would have established an ontological Pythagoreanism. This line of argument will probably appeal to some thinkers oriented towards mathematics. However, we are not aware of any contemporary philosopher who actually advocates a strong Pythagoreanism, and there are weighty arguments against this position. First of all, there are qualitative aspects of the world that resist a mathematical-scientific description. The reader should think here of the smell of a rose, not of the physiological process of smelling, but of what it is like for him to smell the scent of a rose. Such peculiar qualities are irreducible components of the world. Even if we disregard such aspects of the world, a strong Pythagoreanism is still problematic. In the next chapter, we will learn arguments that suggest that the objects of mathematics are not concrete, but abstract. If this is correct, an ontological identification of concrete objects of the world with mathematical objects is ruled out. Even if a scientific theory is mathematically formulated this does not mean that the objects of the theory are mathematical objects. Scientific theories are founded on observations and experiments, i.e., on experience, and therefore refer to the empirical world. Mathematical theories do not seem to be based on experience and therefore do not seem to refer to the empirical world. For epistemological reasons, the identification of objects of the natural sciences with mathematical objects is therefore questionable. We will discuss this point in more detail in Chap. 4. Other monistic ontologies than the Pythagorean one, which only assume the existence of one kind of thing, also seem problematic to us. The claims *Everything is physical* or *Everything is mental* like the claim *Everything is mathematical*, involve considerable philosophical difficulties. This topic will occupy us especially in Chaps. 5, 8 and 12.

Here we want to discuss a weaker variant of Pythagoreanism that does not presuppose a monistic ontology.

---

[7] See for example Barrow (2007).

2.3   Mathematics as a Foundation of Science

A weak or scientific-theoretical Pythagoreanism in philosophy is characterised by the following principle:

Mathematics is a foundation of the empirical sciences.

Had the Pythagoreans known our modern terminology, they would probably have been willing to defend this thesis. We hope that our nomenclature appears acceptable with this.

In all contemporary empirical sciences, there are numerous mathematically formulated models.[8] The mathematisation of the natural sciences goes back to the Italian polymath Galileo Galilei (1564–1642), for whom mathematics was the *language of nature*. An early success of this mathematisation are the laws of planetary motion by Johannes Kepler (1571–1630). Some further highlights of mathematical physics are classical mechanics, thermodynamics, electrodynamics, the theories of relativity and quantum mechanics. All these theories contain a sophisticated mathematical formalism.[9] The value of mathematical models in physics has not been lost on scientists from other disciplines. After physics, the process of mathematisation also took hold in chemistry and biology. For example, there are now areas of work in biology such as mathematical evolutionary theory, biosignal analysis or bioinformatics. Even in the social sciences, psychology and linguistics, we now find numerous mathematically formulated models. Scientific practice thus advocates for a weak Pythagoreanism.

A significant approach of modern empirical sciences to the world consists of counting and measuring. Investigations and experiments provide us with quantitative data, ultimately numbers. The formulation of models in the language of mathematics is therefore close at hand. To examine our models and make predictions, we use methods provided by mathematics. If empirical sciences want to make precise quantitative predictions, the use of mathematical models and methods is evidently indispensable. We can predict without mathematics that an apple that detaches from a tree will fall to the ground. To say how long the fall of the apple takes, we need mathematics. In this sense, we can fully agree with the thesis of the scientific-theoretical Pythagoreanism that mathematics is a foundation of empirical sciences. Nevertheless, we maintain a fundamental scientific-theoretical distinction of mathematics from empirical sciences, mathematics does not require empiricism. We will discuss this fact in more detail elsewhere.[10]

Some scientists assume an *Unreasonable Effectiveness of Mathematics in the Natural Sciences*, especially in physics. Mathematical models and methods are

---

[8] We will not present a theory of mathematical model building in the empirical sciences here and instead refer to Ludwig (1985), for example.

[9] See for example Kuhn (2016) for a history of physics.

[10] See especially Chap. 4.

supposed to be more suitable as a basis for our understanding of natural events than would be expected. The phrase *Unreasonable Effectiveness of Mathematics* comes from the Hungarian-American physicist Eugene Wigner (1902–1995), but a similar sentiment can also be found in Albert Einstein (1879–1955), Paul Dirac (1902–1984) and others.[11] If the mathematical models in the natural sciences were indeed obvious and at the same time of great predictive power, this could be of philosophical significance and used as a justification for a stronger Pythagoreanism. We doubt that this is the case, and are rather surprised at how inefficient our mathematical models of natural processes still are.

We are not aware of any mathematical model of the evolution of the universe that predicts the expansion rate of the universe in different directions or the distribution of the temperature of the cosmic background radiation that we measure. We also do not know of any mathematical model of the formation of our solar system that fully explains the number, mass and orbits of the planets in it. Even more so, we do not have a mathematical model of biological evolution that determines basic properties of species (weight, size, life expectancy, mobility, etc.) and their distribution on Earth. The long-term future of the universe, the solar system and biological species is largely in the dark.[12] Even predicting the weather (air pressure, temperature, wind, precipitation) at a location in a few weeks is not possible for us. Perhaps it is too much to ask in all these examples that mathematical models can make precise predictions. Unfortunately, we also lack sound stochastic models that allow us to estimate the probabilities of different scenarios. Natural processes do not seem to make it overly easy for us in terms of mathematical modelling and quantitative prediction.

Even if there is little to suggest a strong Pythagoreanism in the philosophy of mathematics, weak Pythagoreanism is nevertheless a foundation of the scientific worldview. The application of mathematics is useful for formulating models of the quantitative aspects of the world. It is even indispensable, if we want to make precise quantitative predictions. However, this does not imply that mathematics is a natural or empirical science.

---

[11] See Wigner (1960) for this.

[12] The latter even depends on the behaviour of the human species, which is hardly predictable even in the short term.

Platonism

3

Contents

3.1   Historical Background

Platonism is a metaphysical and epistemological concept, which goes back to the ancient Greek philosopher Plato (428–348 BC) from Athens. Plato was a student of Socrates (469–399 BC) and teacher of Aristotle and is one of the most significant figures in European intellectual history. Plato's work is extensive, original and diverse. In addition to questions of metaphysics and epistemology, it also deals with topics of ethics, aesthetics, anthropology, political theory and cosmology.[1] In addition to his work as a writer, Plato founded an academy. This was the oldest and longest-lasting institutional school of philosophy in ancient Greece. Through the students of the Academy, Plato's teachings spread and became a foundation of ancient civilization. In the Roman Empire, Platonism gained influence through the philosophical school of Plotinus (205–270) in Rome and thus became a dominant doctrine in the philosophy of late antiquity. In the Christian Middle Ages, only a few of Plato's writings were known. An indirect influence of Platonic

---

[1] A complete edition of Plato's works in German translation can be found in Loewenthal (2014). In addition, the Reclam series offers selected works of Plato in the original text with German translation.

J. Neunhäuserer, *Introduction to the Philosophy of Mathematics*,
Mathematics Study Resources 22, https://doi.org/10.1007/978-3-662-72179-7_3

teaching can be mainly demonstrated through Aristotelian philosophy. Only in the Italian Renaissance were Plato's works translated in their entirety from Greek into Latin, commented on and thus influenced humanism.[2] In the philosophy of the Enlightenment, Plato seems to play a subordinate role, only in German Idealism we find, for example, in Hegel (1770–1831) and Schelling (1775–1854) a renewed reception of Plato.

The philosophy of the modern era is generally rather sceptical towards metaphysical concepts, yet Platonism in the philosophy of mathematics, as we will see, gains new significance. In particular, Plato's theory of ideas, which forms the core of his metaphysics and epistemology, is of interest to the philosophy of mathematics. We therefore provide a compact introduction here.[3]

An idea in the sense of Plato is a bodiless, timeless, simple and perfect being. Ideas are not accessible through sensory perception, as they are not located in physical space, and they are not mental events, since they are timeless. The changeable objects in physical space are images of ideas, and thus a lower being, which is derived from the ideas as archetypes. Physical objects have a share in ideas, in that they exhibit the essence of ideas incompletely and imperfectly and are involved in them in this sense. Conversely, ideas allow physical objects certain aspects of their essence, insofar as physical reality allows this. Ideas are present in physical objects in this way. Every physical object has a share in several ideas and each idea allows a multitude of physical objects to share in its essence. Consider, for example, a tall fir tree. This has a share in the ideas fir, tree, plant, living being, space-time object as well as the ideas of height and unity. According to Plato, ideas are ordered by the degree of their generality, the more general is higher-ranking and includes the more specific, the specific has a share in the general. Mathematical objects in their proper form are for Plato ideas and thus bodiless, timeless and unchangeable. A drawn geometric object represents an inaccurate image of the idea of the geometric object, which is the archetype of the object. Thus, a circle drawn with a compass is not the actual object of geometry, it is the idea of the circle, with which geometry deals. Similarly, behind each individual number hides an idea corresponding to it, one can say in the sense of Plato that behind the number one hides the idea of unity, behind the number two stands the idea of duality, behind the number three the idea of triad, and so on. The idea of unity represents the origin and the principle of order not only of numbers, but of the diversity of all ideas. It is occasionally identified by Plato with the idea of the good. The good is for Plato the highest idea, to which all other ideas owe their being, through their participation in the good other ideas are good.

The theory of ideas outlined here by Plato can be described as monistic and objective idealism. In addition to this metaphysics, Plato also develops an

---

[2] The translation of Plato into Latin was done by the Renaissance humanist and philosopher Marsilio Ficino (1433–1499), see Blum et al. (1993).

[3] Plato develops the theory of ideas mainly in the works Phaedo, Republic, Phaedrus and Parmenides, which belong to the middle dialogues.

epistemology, which tries to clarify in what way we have access to the ideas. As said, the Platonic ideas as timeless and incorporeal entities are withdrawn from sensory perception. Plato describes the process of essence viewing, a kind of non-sensory perception, in which we directly grasp an idea. The philosophical conversation serves as a preparation for the essence viewing, in which we develop correct opinions about ideas. This conversation triggers a process of knowledge, at the end of which the original idea stands directly before our eyes. The idea viewing is closely linked with Plato's doctrine of the transmigration of souls, according to which the soul is immortal and successively inhabits different bodies. Between two earthly lives, the soul is bodiless and can recognise the ideas undisturbed and error-free. Through the connection with the body, the abilities of the soul are restricted. With some preparation and effort, the soul can, stimulated by sensory impressions, remember the underlying ideas again. The recollection of the immortal soul is Plato's explanation for the possibility of idea viewing.

Although Plato's theory of ideas and epistemology in relation to nature, resp. physical reality, is rarely represented in the modern era, Platonism in the philosophy of mathematics remains current, both in metaphysical and in epistemological terms. The view that the objects of mathematics are ideas in the sense of Plato and that we directly intuitively grasp basic properties of these objects, is still widespread among mathematicians.

3.2   The Platonic Position

A modern Platonism in the philosophy of mathematics is based on the following three principles (see Fig. 3.1):

1. Mathematical objects such as sets, numbers, relations, functions etc. exist.
2. Mathematical objects are independent of mental processes.
3. Mathematical objects are abstract.

If principles one and two are asserted, one speaks of a realistic philosophy of mathematics. Platonism in the philosophy of mathematics can therefore also be referred to as Platonic realism. We will discuss a non-Platonic realism in the philosophy of mathematics in Chap. 4 and in Chap. 10 on naturalism in the philosophy of mathematics. Starting from Plato's theory of ideas, it is to be suspected that Plato would actually accept the principles of a modern Platonism in the philosophy of mathematics, an assumption that justifies our terminology.

Although the three principles mentioned appear clear and simple at first glance, they can be understood and explained in different ways.

First, there is the ontological question of what it means for a mathematical object to exist. A crucial component of modern mathematics is the assertion of existence and its proof. For instance, the Fundamental Theorem of Arithmetic states that for every natural number n greater than 1, there exist prime numbers whose product is n. The Fundamental Theorem of Algebra asserts that for every algebraic equation

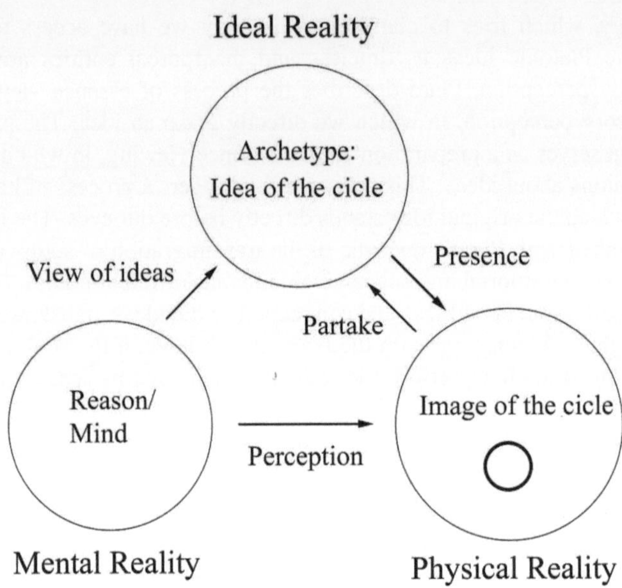

Fig. 3.1 Scheme of the theory of ideas

of degree n, exactly n solutions exist in the complex numbers. Existence assertions in modern mathematics claim the existence of one or more elements in a given set, which satisfy a predicate or transform a propositional form into a true statement. The concept of existence is not used as a predicate here, but as a quantifier, which binds variables in propositional forms and turns them into statements, see also Chap. 6 on Logicism. The proof of an existence assertion generally consists of a deduction of the statement using the rules of predicate logic from the axioms of set theory. Every axiomatic set theory contains axioms, which assert the existence of certain sets.[4] Fundamental to this is the assertion of the existence of the empty set and an inductive set. The second assertion allows in particular the existence of infinite sets, such as the natural numbers. Mathematics itself does not tell us why the axioms of set theory are true and in what sense the sets, whose existence is demanded, exist. As mathematicians, we usually just assume an axiom system and do not question it. It is a task of the philosophy of mathematics to do just that. Platonism assumes that the existence of mathematical objects is objective. This means, that the objects of mathematics would exist even if there were no mental processes. In particular, the objects of mathematics would still exist if our rational efforts, to make statements about these objects and to justify these statements, were to disappear. Even if there were never any kind of cognising being, yes even if not a

---

[4] In contemporary mathematics, the axiom system of Ernst Zermelo (1871–1953) and Abraham Fraenkel (1891–1865) is predominant. For an introduction to axiomatic set theory, we refer to the appendix and Deiser (2010).

single thought existed, the objects of mathematics would exist. We can understand this Platonic realism in relation to mathematical objects in analogy to a realism in relation to physical objects. Just as the Earth, the solar system and the physical universe would exist even if there were no humans or other cognising beings, so too would sets, relations, functions and numbers exist. This analogy should be treated with caution. A modern Platonism in the sense of the great Austrian-American logician Kurt Gödel (1906–1978) assumes that not only do the objects of mathematics exist independently of mental processes, but also, that the properties of mathematical objects are independent of such processes.[5] Even if no mental processes existed or these were completely different, the properties of mathematical objects would remain the same. Under this condition, Platonic realism can be easily understood in analogy to physical realism. The so-called neo-Fregean philosophy of mathematics is also Platonic in the sense of the three principles mentioned above, but asserts that the ontological structure of mathematical reality is determined by our mathematical cognition process.[6] Our cognition process therefore determines how mathematical reality is structured into objects, properties and relations, while the actual subject of mathematics remains independent of this structuring. This is a minimal Platonic realism in the philosophy of mathematics, which can hardly be understood in analogy to realism in relation to macroscopic physical objects.

With the determination of mathematical objects as abstract, Platonic realism is clearly distinguished from physical realism. Physical objects are usually defined as spatio-temporal and we expect from abstract objects in any case, that they are not spatio-temporal. An object that is not spatio-temporal, can however only have spatial but no temporal properties or only temporal but no spatial properties. Further central determinations of the concept of an abstract object exclude this possibility. An abstract object is timeless and unchangeable, there are no changes in the intrinsic properties of such an object. By intrinsic properties, we mean those properties that the object has independently of its relations to other objects. Furthermore, abstract objects are acausal, they have no causal influence on other objects and are not causally influenced by other objects . It is usually assumed that causal relations are relations that exist between objects of spatio-temporal reality and only between such objects. Lastly, it is sometimes assumed of abstract objects that they necessarily exist. It would not be possible for these objects not to exist. Whether the last condition necessarily applies to every abstract object, is disputed. That mathematical objects are paradigmatic examples of abstract objects with the properties of timelessness and acausality, is however undisputed in the Platonic philosophy of mathematics.

---

[5] See Gödel (1940) and Parsons (1995) for a discussion of Gödel's Platonism.

[6] The British philosophers Bob Hale (1945–) and Crispin Wright (1942–) are the main representatives of this direction, see Hale and Crispin (2001). As with the German mathematician Gottlob Frege (1848–1925), Hale and Wright's philosophy of mathematics includes both Platonic considerations and logicist approaches, compare to this Chap. 7.

So far, we have explained the concept of Platonism in the philosophy of mathematics. The question now arises as to what reasons speak in favour of Platonism.

3.3   Arguments for Platonism

The first argument for Platonism in the philosophy of mathematics is often overlooked by philosophers, even though it provides a good reason to at least seriously consider Platonism.

(1) A large majority of mathematicians are convinced that the objects they deal with exist independently of mental activity and are abstract.
(2) A large majority of representatives of a mature science are not mistaken about the fundamental properties of the objects they deal with.

From (1) and (2) it follows: Platonism in the philosophy of mathematics is true.

The first premise of the argument is an empirical claim. We assume that a survey among mathematicians in academic research and teaching would show that it is indeed true. Our reasons for this belief are manifold. Firstly, the literature often states that Platonism is the standard position of mathematicians, and a non-representative survey among mathematicians in our acquaintance confirms this. In a survey of 931 philosophers by PhilPapers, Platonism in relation to abstract objects is the most commonly held position,[7] and among mathematicians, Platonism seems to be even more widespread than among philosophers. Finally, we would like to point out that Paul Erdös (1913–1996), one of the most influential mathematicians of the twentieth century, held Platonic views.[8]

The second assumption of the argument is at least problematic, if not outright false. The history of science shows that scientists can be mistaken, even about the fundamental properties of the objects they deal with. Physicists assumed until the end of the nineteenth century that light is a wave that propagates in an ether. We now know that the luminiferous ether does not exist and that light has both wave and particle properties. The assumption that such errors no longer occur in a mature science is undoubtedly bold. It seems difficult to specify when a science has reached the status that the majority of scientists do not hold false views on fundamental philosophical questions. Nevertheless, we do not consider the second assumption of the argument to be irrational, assuming an epistemological optimism. Therefore, in our view, the argument provides a debatable indication for Platonism in the philosophy of mathematics.

The second argument for Platonism in the philosophy of mathematics consists of several parts, which are intended to individually justify the three principles

---

[7] See https://philpapers.org/surveys/.
[8] See the Erdös biography by Hoffmann (1999).

of Platonism. The argument essentially goes back to the German logician and philosopher Gottlob Frege (1848–1925) and finds support among philosophers.[9] We present the argument here not in Frege's style, but in a contemporary version. The argument for the existence of mathematical objects is as follows:

(1) Singular terms that refer to mathematical objects, such as numbers, appear in simple and true statements.
(2) It is only possible for simple statements with singular terms to be true if the objects to which the singular terms refer exist.

From (1) and (2) it follows: Mathematical objects exist.

A singular term, according to linguistics, is an expression that denotes exactly one object; in contrast, a predicate is an expression that stands for a concept, under which several objects or no object at all can fall. Examples of singular terms that refer to mathematical objects are *one, two, three*, etc. It is hard to doubt that such terms appear in true statements, we usually assume that *two is equal to one plus one* or *one is equal to two minus one* are true statements. This is precisely the meaning of the first premise of the argument. It should be noted that there is a current in contemporary philosophy of mathematics that tries to undermine this premise, see Sect. 13.3 on fictionalism. The objects to which singular terms refer belong to the ontological category of objects. In contrast are the ontological categories of properties. It does not seem sensible to assume that objects that do not exist have properties. Simple statements with singular terms attribute properties to objects. These statements cannot be true if the objects being talked about do not exist. This is a justification of the second premise of the argument, which is in the spirit of Gottlob Frege.

We do not necessarily have a singular term for each natural number in a natural language, and we do not have such a term for each real number in any language. However, it is possible to give a singular term for each individual given natural or real number that denotes the number. The argument therefore justifies the existence of individual numbers. In modern mathematics, however, not individual numbers, but sets and in particular sets of numbers are considered. There are statements about such sets, such as *the set of natural numbers is the union of the set of even and the set of odd natural numbers*, the truth of which is hard to doubt. If we now want to apply the argument, the question arises whether *set of natural numbers* is a singular term that denotes exactly one object. A set is in particular a class, i.e. a collection of objects, and the *set of natural numbers* is a so-called class expression. Class expressions are singular if they refer to a uniquely determined class. In modern mathematics, starting from an axiomatic set theory, the uniqueness of classes such as that of natural numbers is demonstrated. Formally, the natural numbers are uniquely determined as the intersection of all inductively ordered sets in an axiomatic set

---

[9] See Frege (1884). Besides his Platonic argumentation, Gottlob Frege is a main representative of logicism in his later works, see Chap. 7.

theory.[10] If we accept this, the argument can be applied not only to individual numbers, but also to individual sets, demonstrating their existence.

An argument for the independence of mathematical objects from mental processes is as follows:

(1) The statements of mathematics are objective.
(2) The singular terms of objective statements refer to objects that are independent of mental processes.

From (1) and (2) it follows: The objects of mathematics are independent of mental processes.

That a statement is objective means that the truth value of the statement is independent of mental processes. Overlooking errors and misunderstandings, all mathematicians establish the same truth value for a mathematical statement through proofs. This is a strong indication for the first premise of the argument. If the objects to which singular terms in statements refer depend on mental processes, the properties of such objects and thus the truth value of statements about these objects depend on mental processes. This is an argument for the second premise of the argument. A rationalistic philosophy of mathematics will deny the conclusion and at least the second premise of this argument, see Chap. 3 on Rationalism for this.

Starting from the last argument, we get another argument that the objects of mathematics are abstract:

(1) Mathematical objects are connected by true identity statements with mathematical concepts.
(2) No physical object is connected in this way with mathematical concepts.

From (1) and (2) it follows: (3) The objects of mathematics are not physical.

(4) Objects that are independent of mental processes and not physical are abstract.

From (3) and (4) it follows: The objects of mathematics are abstract.

Consider a mathematical concept like *The number of prime numbers less than ten* or *The positive solution of the quadratic equation* $x^2 - 1 = 0$. These concepts refer to mathematical objects. The identity statements *The number of prime numbers less than ten is four* and *The positive solution of the quadratic equation* $x^2 - 1 = 0$ *is 1* are true. These identities show the special way in which mathematical concepts are connected with mathematical objects; they only apply to this type of objects. For every mathematical object, such a mathematical concept and a corresponding identity statement can probably be constructed. The claim of the identity of a mathematical concept with a physical object seems implausible, *The number of*

---

[10] See Deiser (2010) for more on this.

prime numbers less than ten apparently cannot be related to a physical object. We therefore arrive at the difficult to dispute conclusion, that the objects of mathematics are not physical. The objects that become temporal, changeable and possibly causally effective, are to our knowledge either mental or physical. An object that is independent of mental processes and not physical, is thus timeless, unchangeable and acausal, hence abstract.

So far, we have outlined an argumentation line inspired by Gottlob Frege and oriented towards linguistics, which suggests that mathematical objects exist independently of mental processes and are abstract. We will subsequently present a contemporary argument for realism in the philosophy of mathematics and another argument for the abstractness of mathematical objects. The first argument is as follows:

(1) We should accept as real the existence of all those objects that are indispensable in our best scientific theories.
(2) Mathematical objects are indispensable in our best scientific theories.

From (1) and (2) it follows: We should accept the existence of the objects of mathematics as part of reality.

This argument goes back to the American philosopher and logician Willard Quine (1908–2000) and was extensively presented by the American philosopher Hilary Putnam (1926–2016).[11] The second premise of the argument is undisputed: No one doubts that objects of mathematics such as numbers, relations, functions etc. are at least indispensable in the natural sciences, see also Sect. 2.3. The first premise of the argument requires further explanation and justification. Behind the claim that the objects indispensable in scientific theories are real, is the view, that scientific theories represent our best access to reality and will be able to describe the world comprehensively and correctly. The ontological question, what kind of objects exist, should therefore be answered by referring to scientific theories; an unscientific way to determine what exists is excluded. This metaphysical position is referred to as naturalism or scientism, see also Chap. 11 on Naturalism in the Philosophy of Mathematics.[12] Naturalism denies the legitimacy of metaphysics as an independent discipline separate from the sciences. Although naturalism is very widespread in scientific circles, it and thus also the first premise of the argument are philosophically at least problematic. As we had already noted at the beginning of the section, scientific theories can be false. They can make the existence of objects appear indispensable, which no longer occur in a further developed theory. We refer here again to the assumption of the luminiferous aether in physics. Furthermore, metaphysics possesses a valuable pre-scientific methodology. People, and scientists in particular, have certain basic intuitions; metaphysics explicates these and derives

---

[11] See Quine (1963, 1981) and Putnam (1971) for more on this.

[12] In Keil and Schnädelbach (2000) there is a collection of contributions from naturalists and anti-naturalists in philosophy.

arguments in a natural language for or against a metaphysical position from them. These philosophical arguments can argue for the existence of an object that does not appear in our best scientific theories; or they can argue against the existence of an object that seems indispensable in our best scientific theories. Ignoring these arguments is a sign of scientific arrogance. Despite this vehement criticism of naturalism, we believe that the above argument provides a tolerable justification for realism in the philosophy of mathematics. The fact that mathematical objects are indispensable in our best scientific theories, which describe reality independently of us, can be taken as an indication that mathematical objects exist independently of us.

Assuming a realism regarding mathematical objects, we have the following argument for Platonism, which belongs to the folklore in the philosophy of mathematics:

(1) Concrete objects that exist independently of mental processes are at a place at a time.
(2) Mathematical objects are not at a place at a time.

From (1) and (2) it follows: The objects of mathematics are not concrete, but abstract.

We assume that concrete objects that exist independently of mental processes are physical. Physical objects are at a place at a time, or to specify this in appreciation of the results of modern physics: Physical objects like elementary particles are associated with a region in space-time. When we imagine concrete objects, like ghosts and gods, that exist independently of mental processes and are not physical, we also imagine these objects at a place at a time. Concrete objects that exist independently of mental processes and are not at a place at a time seem unimaginable to us. This is not an incontrovertible proof, but a strong indication of the validity of the first premise of the argument. For arithmetic objects like numbers, the second premise of the argument is obvious; for example, it makes no sense to ask where the nine is at what time. One can say, that geometric objects like lines and circles have a place in a vector space where they are located, the second premise of the argument still applies to these objects as well. Firstly, the question of when a geometric object is at a place in a vector space is meaningless; these objects are therefore not at a place at a time. Secondly, the talk of the place of a geometric object in modern mathematics is not to be taken literally. We only specify for a geometric object which subset it is in a superset. Sets, in turn, do not have a place where they are at a time. For these reasons, we are convinced that the objects of mathematics certainly do not exist at a place at a time.

The last argument of this section has, we hope, made it clear that it is difficult to doubt a Platonism in relation to the objects of mathematics if one starts from a realism in the philosophy of mathematics. Realism in the philosophy of mathematics seems to be much more controversial than the abstractness of mathematical objects.

3.4 The Metaphysical Gap

Humans, as a biological life form, are undoubtedly temporal beings. The objects of mathematics, such as numbers, according to a Platonic philosophy are timeless and exist independently of humanity. There is therefore a fundamental metaphysical gap between humans and mathematical objects. It seems mysterious how we are able to bridge this gap, to gain knowledge about mathematical objects or even just to refer to mathematical objects. The first problem can be described as an epistemic or epistemological challenge and the second problem as a referential challenge for Platonism in the philosophy of mathematics. The Platonic claim that mathematical objects are abstract excludes the possibility that we can refer to mathematical objects unambiguously and gain knowledge about these objects through a causal connection. In contemporary philosophy, causal relations are usually used to explain how we refer to physical objects and gather basic knowledge in the empirical sciences. The perception of objects is based on stimuli from the environment or the body's interior, which are processed by the nervous system and thus provide us with information. A causal reference and theory of knowledge is not applicable to mathematical objects, should these objects be abstract.

We present here two ways that a Platonist can take to meet the referential and the epistemological challenges. The first way, which goes back to Plato's original theory of knowledge, we want to call intuitive Platonism in the philosophy of mathematics.[13]

A representative of intuitive Platonism claims that we are the kind of beings who have an immediate intuitive access to mathematical objects. This ability allows us to refer unambiguously to sets, relations, functions and numbers and to immediately see the truth of simple basic mathematical statements. In particular, this should be the case for the basic axioms of mathematics, as we find them in an axiomatic set theory. Assuming this, neither the reference to nor the gain of knowledge about mathematical objects, that exist abstractly and independently of us, seem mysterious. Intuitive Platonism can leave it at this explanation or try to further explain where our ability to directly intuitively grasp mathematical objects comes from. Such an explanation requires strong metaphysical claims. In any case, one must assume that we are more than a biological life form, as we as such only causally interact with objects that exist independently of us. The assumption of an immortal human soul, as found in Plato and in all world religions, is just one possibility here. One might assert the existence of a timeless consciousness that cannot be reduced to the biological basis, or of a collective, timeless cultural identity of man. However, each of these assumptions is highly speculative.

The second way to meet the epistemic and referential challenges, we want to call theoretical Platonism. In the English-speaking literature, the term full-blooded

---

[13] For example, Kurt Gödel advocated this position, see Gödel (1940) and Parsons (1995).

Platonism is usually used in this context, which we find misleading.[14] This approach is based on two theses:

1. The reference of mathematical theory to the world of mathematical objects is schematic.
2. The world of mathematics contains objects that relate to each other in every logically possible way.

These initial theses require explanation. We call a domain of objects a model for a theory, if all statements of the theory are true with respect to the model. If we assign objects from a domain to the singular terms of the theory in a suitable way, the statements of the theory correctly describe the model of the theory. The reference of the singular terms of a theory to the objects of a model of the theory is called schematic. For a theory, there are generally several models, and we have the freedom to choose one of the models to which the theory refers. A singular term in a theory thus refers to an object in each model of the theory, this is what is meant by schematic reference. If we have formulated a consistent mathematical theory, according to the second thesis, it has at least one model in the world of mathematics. We are thus claiming here that the mathematical world is so large that it contains at least one part for each consistent mathematical theory that represents a model of the theory. The reference of the mathematical theory to its models in the mathematical world is schematic in the sense defined above.

The schematic reference described in this way does not require a causal relation between us, when we set up a mathematical theory, and the mathematical objects to which we refer. This is indeed a way to meet the referential challenge. The epistemological challenge to Platonism is also averted by theoretical Platonism. Through consistent mathematical theories, we gain objective knowledge about a part of the mathematical world, as there is a model in the world of mathematical objects for each of these theories to which our theory refers.

Although theoretical Platonism will appear attractive to those who share Platonic basic intuitions, it has a number of weaknesses. The fact that every consistent mathematical theory we can formulate refers to objects that exist independently of us is a very strong and hardly justifiable metaphysical claim. A proponent of intuitive Platonism will probably deny that every possible consistent axiom system of mathematics refers to the Platonic reality. He will rather want to reject most axiom systems on the basis of immediate intuitive insight and to distinguish others as appropriate. Furthermore, the concept of schematic reference has a certain arbitrariness. Consider, for example, the Peano arithmetic of natural numbers, there are models of this theory that differ significantly.[15] Theoretical Platonism does not allow us to designate one of these models as natural numbers, as the theory refers to each of these models. Finally, the concept of the consistency of a mathematical

---

[14] This position is represented, for example, in Balaguer (1998) and Shapiro (1997).

[15] See Kaye (1991) for this.

theory, which theoretical Platonism presupposes, is problematic. We will go into this in detail in Chap. 9 and only note here that the consistency of a sufficiently strong mathematical axiom system cannot be proven within such a system. In the next section, we will see that there are axiom systems of mathematics that are equivalent in terms of their consistency but imply very different statements in classical areas of mathematics. A theoretical Platonist must assume that each of these systems refers to a part of mathematical reality.

3.5   Platonism and the Continuum Hypothesis

One of the great problems in the foundations of mathematics is the so-called Continuum Hypothesis. The question is whether there is a set that is larger than the natural numbers in the set-theoretical sense, but smaller than the real numbers. More precisely, it is about whether there is a subset of the real numbers that contains the natural numbers, but can neither be mapped one-to-one onto the real numbers nor onto the natural numbers. The Continuum Hypothesis states that such a set does not exist. However, it has been shown that this question is undecidable in conventional set theory. An ordinary axiom system of set theory, such as the Zermelo-Fraenkel system with the axiom of choice, implies no contradictions if and only if it, extended by the Continuum Hypothesis, implies no contradictions. However, it also implies no contradictions exactly when, extended by the negation of the continuum hypothesis, it implies no contradictions.[16] So we can add the continuum hypothesis or its negation to the usual axioms of set theory without contradiction. So far, it seems to be our decision whether there is a set that is larger than the natural numbers in the set-theoretic sense, but smaller than the real numbers. Depending on whether we accept the continuum hypothesis or its negation, we obtain isolated statements of geometry, analysis and topology that contradict each other.[17]

This is a serious problem for a Platonism with respect to the real numbers and thus for a Platonism with respect to a large part of modern mathematics, in which the real numbers are used. If we can actually decide on the existence of certain subsets of the real numbers, the real numbers do not exist independently of our mental processes, so Platonism with respect to the real numbers would be wrong. The same applies to all areas of mathematics, such as analysis or modern geometry and algebra, in which the real numbers are used.

We see three ways in which a Platonist can deal with this problem.

The first possibility is to assume, in the sense of intuitive Platonism, that a simple and immediately comprehensible axiom of set theory is waiting to be discovered by

---

[16] See Kunen (1980) for an introduction to the proofs of the independence of the continuum hypothesis. The axiom of choice states that there is a function for a family of sets that selects an element from each of the sets. This axiom is indispensable in a significant part of modern mathematics, see also Sect. 7.5.

[17] See for example Sierpinski (1965) and Erdös (1953–1954).

humanity and this axiom decides the continuum hypothesis. This would mean that we do not yet understand the concept of a set correctly and we will only know when this is the case, what size sets exist between the natural and real numbers. Against this Platonic perspective is the fact that the continuum hypothesis has been known since the foundation of set theory by Georg Cantor (1845–1918) and many great mathematicians have thought about the continuum hypothesis without being able to suggest a simple axiom, which solves the problem.[18] It is still possible that the Platonic hope will be fulfilled.

The second possibility of dealing with the continuum hypothesis is to assume, in the sense of theoretical Platonism, that at least two universes of mathematical objects exist, with the continuum hypothesis holding in one universe and its negation being true in another. There would therefore be at least two sets of real numbers, and depending on which mathematical universe we are talking about, we would be referring to different sets of real numbers. These sets would differ significantly in their properties and especially in relation to their subsets.[19] This solution reminds a little of the many-worlds interpretation of quantum mechanics, which claims that the universe splits through a measurement on a quantum system and all measurements possible according to the theory actually occur.[20] Such a solution to the problem of the continuum hypothesis seems to us to be philosophically unattractive, as it contradicts the principle of ontological parsimony. This principle, named after the English philosopher Ockham (1288–1347), is also called *Ockham's razor* and states that we should try to assume the existence of as few objects as possible to explain the phenomena we are dealing with.[21] A many-worlds interpretation of mathematics multiplies the world of mathematical objects whose existence we must assume. If there is a philosophical alternative, this should be avoided.

The third possibility is to limit Platonism in the philosophy of mathematics to the natural numbers and to concede, that the mathematics that goes beyond these is actually dependent on our mental processes. The German mathematician Leopold Kronecker (1823–1891) meant in this sense *God made the natural numbers, all else is the work of man*. A restricted Platonism can be specified in the following way: The objects whose existence follows from the axioms of a basic set theory exist in the Platonic sense and are independent of our cognitive process. The objects whose existence is inferred from further axioms such as the axiom of choice do not exist independently of mental processes. However, it seems arbitrary where the boundary between the basic and an extended set theory is drawn. We could limit Platonism, for

---

[18] At the beginning of the 2000s, the American mathematician William Hugh Woodin (1955–) argued against the validity of the continuum hypothesis based on the existence of large cardinal numbers, see Woodin (2001). However, he later revised his view and the continuum hypothesis is still open.

[19] The many-universes interpretation of set theory is currently being intensively discussed, see for example Hamkins (2012).

[20] See Barrett and Byrne (2012).

[21] See for example Hübener (1983).

example, to finite sets and assume that even the existence of an inductively ordered set and thus the set of natural numbers is not to be understood in the Platonic sense. See also Chap. 8 on intuitionism. In any case, mathematics would deal with objects that differ fundamentally in their mode of being. If we take a naive perspective on mathematics as a whole, this does not seem to be the case. Mathematics presents itself to us uniformly in terms of its subject area. The three strategies described here for how a Platonist can deal with the problem of the continuum hypothesis are certainly worth considering, how convincing the alternatives are, the reader may judge.

Rationalism

4

Contents

4.1 Introduction

The basic assumption of rationalism is that we, as rational beings, possess a priori knowledge, that is independent of any experience and does not require justification through experience. This knowledge is either directly given to our reason or it is uncovered by the activity of pure reason, without recourse to experience. The activity of pure reason is usually described as deduction, i.e., our reason draws valid conclusions from given premises. Rationalism is an epistemological position in philosophy. Epistemology is the part of philosophy that deals with the question of what knowledge is, how we can acquire knowledge and how statements can be justified.

From the basic assumption of rationalism, it does not follow that all our knowledge is independent of experience. However, if rationalism is advocated in relation to a specific subject area of knowledge, this means that all true statements about the objects of the area are justified a priori, i.e., independent of experience. A rationalism in relation to mathematics asserts that we directly perceive the truth of the axioms of mathematical theories and derive mathematical statements from these premises using logical deduction, i.e., logical inference rules. How mathematical axioms are given to us is explained in different ways in rationalism. Either this knowledge is always part of our rational nature, i.e., innate, or it is made accessible to us through an intuitive process of cognition. An intuitive process of cognition can

J. Neunhäuserer, *Introduction to the Philosophy of Mathematics*,
Mathematics Study Resources 22, https://doi.org/10.1007/978-3-662-72179-7_4

be understood in analogy to sensory perception: We see a mathematical fact before our mind's eye. For the rationalist, however, intuition is a cognitive mental process and not part of our sensuality or emotionality.

Rationalism is opposed by empiricism in epistemology. Empiricism asserts that there is no other source of knowledge than experiences, i.e., we can only justify opinions empirically, i.e. through experience. Empiricism can be limited to specific subject areas of knowledge. An empiricism limited to such an area asserts that all true statements about the objects of the area are empirical, i.e., can only be justified through experience. If two subject areas are well distinguished, a rationalism in relation to one area can be reconciled with an empiricism in relation to the other area. If the subject areas, however, are not assumed to be disjoint, empiricism and rationalism contradict each other. Here is an example: If we assume that the objects of mathematics do not occur in nature, an empiricism in relation to nature is compatible with a rationalism in relation to mathematics. However, if we assume that objects of mathematics can be found in nature, empiricism in relation to nature contradicts rationalism in relation to mathematics. The statements of mathematics cannot be both empirical and a priori, i.e., justified only through experience and justified independently of experience.

In the history of philosophy, a philosophical system is called rationalistic, if it puts a priori knowledge and deductive justification at the centre and ignores or devalues empirical knowledge and justifications through experience or devalues them. A system is called empirical if it identifies experience as the only or most important source of our knowledge.[1] Already in antiquity, rationalistic approaches to epistemology can be found in Plato and Aristotle. We have already discussed Plato's philosophy in Chap. 3, his student Aristotle developed with his syllogistics a doctrine of correct reasoning, which can be understood as a precursor of modern logic. The classical rationalistic systems emerge in the early modern period. These are the systems of the French philosopher René Descartes (1596–1650), the Dutch philosopher Baruch de Spinoza (1632–1677) and the German philosopher Gottfried Wilhelm Leibniz (1646–1716). In the history of philosophy, the continental European rationalists are traditionally opposed by the British empiricists John Locke (1632–1704), George Berkeley (1685–1753) and David Hume (1711–1776). Whether the philosophy of the great English natural scientist Isaac Newton (1642–1726) can be described as rationalistic, or whether it represents a variant of empiricism, is disputed. We consider this question in detail in Sect. 4.3. The philosophy of the German philosopher Immanuel Kant (1724–1804), which we will discuss in more detail in Chap. 5, can be read as an attempt to mediate between rationalism and empiricism. This rough historical classification is certainly simplifying. A commonality of the empiricists and rationalists lies in the departure from divine revelation, i.e., the Bible, as the fundamental source of our knowledge. This departure, whether rationalistically or empirically founded, was a necessary

---

[1] For readers interested in the history of philosophy, we recommend Russel (2004) and Störig (2016).

prerequisite for scientific progress in the modern era. On the other hand, the individual rationalistic and empirical systems differ significantly in their principles. A uniform rationalistic or empirical metaphysics is not recognisable.

In Descartes, Leibniz and also in Newton, we find philosophies of mathematics that differ significantly in terms of ontological questions, but all contain a rationalist theory of knowledge with respect to mathematics. We discuss these theories in Sects. 4.2–4.4. Spinoza uses the mathematical method in the form of definitions, principles, theorems and proofs in his main work *Ethica, ordine geometrico demonstrata (1677)* to design a metaphysics, anthropology and ethics in accordance with Euclid's geometry.[2] We cannot discover a systematic design of a philosophy of mathematics in Spinoza. Therefore, we do not go into more detail about this great thinker. The sophisticated philosophy of mathematics of Kant will occupy us in Chap. 5.

4.2   Descartes

The philosophy of René Descartes forms the first original and influential philosophical system of modern times. Descartes is therefore considered the founder of modern philosophy. Bertrand Russell even believed that nothing of comparable importance in the history of philosophy can be found between Aristotle and Descartes. In his main philosophical works *Discours de la Méthode (1637)* and *Meditationes (1642)* Descartes starts from a radical doubt, which is known today as *Cartesian doubt*.[3] Descartes doubts both our sensory perception and our thinking. It could be that a powerful demon systematically deceives us and everything is different than it appears to us or as we think. But if I doubt, it is impossible to doubt that I doubt, so I think and therefore exist. If I did not exist, I could not think and especially not doubt. This leads Descartes to his famous sentence *Cogito ergo sum* (I think, therefore I am). This certainty is the starting point of Descartes' metaphysics, the existence of the *res cogitans*, a thinking substance, is secured. Similar considerations as in Descartes can already be found in the church teacher Augustine of Hippo (354–430) in late antiquity. In contrast to Descartes, for Augustine the thinking subject, which undoubtedly exists, is not the basis of knowledge or the starting point of philosophy. For Descartes, the sentence *Cogito ergo sum* is the prime example of a true sentence and the paradigm of his theory of knowledge. Everything we recognise as clearly and distinctly as this sentence must also be just as certain. Descartes now resorts to arguments of medieval scholasticism, to prove the existence of God, which is hardly compatible with his supposedly radical doubt and strong concept of truth. At least for us, there is no clear and distinct knowledge of a perfect being. For Descartes, the world of bodies exists because God, as a perfect being, is truthful and does not lead us astray like an evil demon. The characteristic property of the

---

[2] See de Spinoza (2020).

[3] The main works of Descartes are Descartes (2018, 2017).

world of bodies is, as we immediately clearly and distinctly recognise, extension. Descartes' metaphysics thus opposes the *res cogitans* with the *res extensa*, an extended substance, The *res cogitans* and *res extensa* are well distinguished, as the thinking substance according to Descartes has no extension. However, there is an interaction between body and mind in humans, which Descartes assumes is mediated by the pineal gland in the brain. The mind is not the cause of a movement of the body for Descartes, but it does have an influence on the direction of the movement. From the perspective of today's brain research and also from the perspective of contemporary physics, Descartes' original position seems untenable. The Cartesian dualism can certainly be formulated in a contemporary way: there is the mental world of cognitive processes and the physical world of space-time processes, with the cognitive processes interacting with certain processes in the brain. This ontological position is rather marginalised in today's international scientific community, but was represented by outstanding scientists such as the brain researcher and Nobel laureate John Eccles (1903–1997) and the famous English mathematician and physicist Roger Penrose (1931–).[4]

After this brief introduction to the Cartesian world of thought, we now come to Descartes' mathematics and philosophy of mathematics. Descartes published only one mathematical work with *La géométrie*.[5] In this work, we find fundamental ideas that led to the development of analytic geometry. In the mathematics before Descartes, geometry was studied separately from arithmetic and algebra. With the application of algebra in geometry, Descartes demonstrated the fruitful connection of the two areas. He thus achieves a partial solution to one of the most famous problems of ancient geometry, which goes back to the mathematician Apollonius of Perga (265–190 BC). The Apollonian problem consists in constructing the circles that touch any three given circles.

In *La géométrie* we find the essential idea of representing points on a line by numbers and vice versa assigning numbers points on a line. Descartes also uses coordinates in his work, i.e. pairs of numbers, to describe points on curves in the plane.[6] The Cartesian coordinate system with two or more orthogonal axes can be traced back to the comments on *La géométrie* by the Dutch mathematician Frans van Schooten (1615–1660), see Fig. 4.1 for this.[7] Furthermore, Descartes introduces a number of symbols, such as the equals sign $=$, the square root symbol $\sqrt{a}$ or the exponential notation a^x, which were of great importance for the development of the symbolic language of mathematics.

In the philosophy of mathematics, Descartes represents a clear rationalist theory of knowledge, coupled with a non-Platonic realism with regard to the objects of

---

[4] See Eccles and Popper (2014) and Penrose (2016) for this.

[5] This work forms an appendix to the *Discours de la Méthode*.

[6] These ideas can also be found in the work of the French mathematician Pierre de Fermat (1607–1665).

[7] We rely here on Burton (2011). It is also reported that the Cartesian coordinate system was known before Descartes. We were unable to verify this claim from sources.

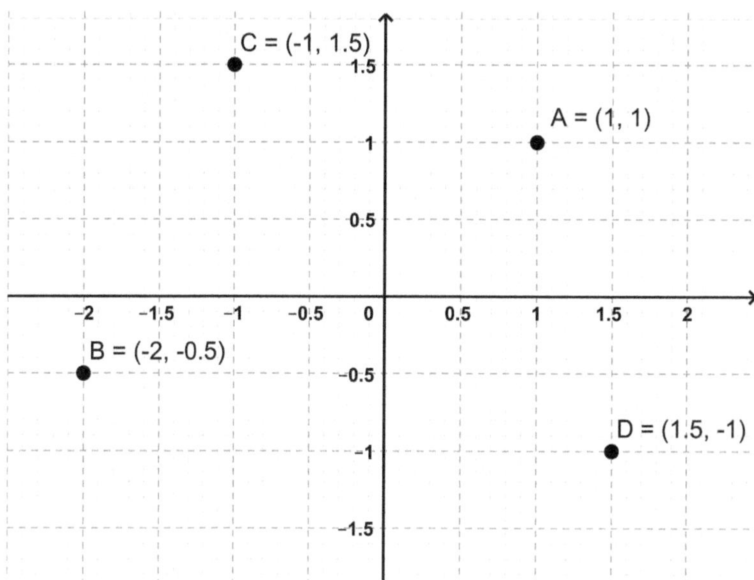

Fig. 4.1 The planar Cartesian coordinate system with four points

mathematics. The method of gaining mathematical knowledge is analytical in nature for Descartes, consisting of intuition and deduction and relying on nothing that is given by experience and thus uncertain. A mathematical intuition for Descartes is an immediate clear and distinct perception of pure and mindful reason, which we cannot doubt. This is simpler and safer than deduction, which according to Descartes determines everything that can necessarily be derived from propositions, which were recognised with certainty. Descartes does not develop a formal logic that specifies when a deduction is conclusive. He limits himself to stating general rules that should determine scientific practice. For Descartes, mathematics is the model of all sciences and should be expanded into a *mathesis universalis*.

The objects of mathematics for Descartes are quantitative properties of the *res extensa*, the world of extended bodies. These exist independently of the *res cogitans*, the world of thoughts. In this sense, Descartes is a realist in the philosophy of mathematics. However, he is not a Platonic realist, as mathematics does not deal with ideas, but quantifiable properties of the material world. According to Descartes, we have an immediate clear and distinct intuition of the extension of material bodies, which allows us to make statements about these bodies independently of experience. Our idea of extension describes the actual extension of material bodies, and this is for Descartes the essential property of the material world. Descartes abstracts from other properties of material bodies, such as colour or solidity, to get to the geometric essence of the *res extensa*, which, as Descartes believes, mathematics deals with.

Descartes' rationalist theory of knowledge seems to us, limited to mathematics, still acceptable today. We accept axioms of mathematical theories based on an immediate intuitive insight into their validity and justify the propositions of mathematical theories through proofs, i.e., deductive derivations using logical inference rules. However, we doubt that pure mathematics provides us with any information about the material world or physical reality. Our fundamental access to physical reality lies in our sensibility. The justification of a statement about physical reality always requires recourse to experiences, such as observations or measurements. One could call this position a minimal empiricism. Even the geometry of physical reality cannot be determined a priori. Descartes only knew Euclidean geometry, which describes flat uncurved space. We know today that there are alternatives to this geometry that describe spaces with curvature. Which geometry applies to physical reality is a question that cannot be decided without observations and measurements. The statements of pure mathematics are, in our opinion, indeed a priori true, their justification comes without recourse to experience. For this reason, mathematics in itself can contain no knowledge about physical reality and in particular no knowledge about the geometry of nature. Of course, we do not deny that mathematical models and methods are very successfully used in empirical sciences. Every mathematically formulated and substantial theory of empirical science contains statements that can only be empirically tested.

4.3   Newton

The Englishman Isaac Newton (1642–1726) is considered one of the most significant natural scientists in the history of science. In his work *Philosophiae Naturalis Principia Mathematica (1687)*, he formulated the three laws of motion of classical mechanics and the law of gravitation.[8] From these laws, he was able to derive, among other things, the three Kepler's laws of planetary motion. To this day, Newton's laws are a central part of classical physics. They allow reliable empirical predictions with certain restrictions on the validity range. In addition to his work on mechanics, Newton's work also includes influential work on optics, acoustics and thermodynamics. In mathematics, Newton developed the calculus, the foundation of analysis, around the same time as Gottfried Wilhelm Leibniz. In Newton's work, the fundamental theorem of calculus is already found, which allows integrals to be determined by reversing differentiation. Furthermore, the classification of planar cubics in algebraic geometry and the generalisation of the binomial theorem for non-integer exponents is attributed to Newton.

Newton's philosophy, based on his scientific work, is often understood as natural philosophy. According to Newton, nature encompasses both passive matter and the immaterial laws of nature that determine the changes in matter. Unlike Descartes, Newton does not advocate an ontological dualism between nature and mind, or between physical and mental reality. However, like Descartes, Newton speculates theologically about the role of God in the world. Nevertheless, in relation to nature,

---

[8] See Newton (2016).

Newton advocates a distinctly empirical theory of knowledge. Observations and experiments are, for him, the fundamental sources of knowledge about nature and the laws of nature. In this sense, Newton can be understood as an empiricist. If we consider Newton's philosophy of mathematics, our impression changes. The basic position of Newton and Descartes in the philosophy of mathematics is identical. For both thinkers, the statements of mathematics are a priori true and refer to physical reality. However, Newton goes a step further in abstraction than Descartes. For Descartes, as mentioned, the extension of bodies was the subject of mathematics and space for Descartes is nothing other than the extension of bodies. For Newton, the extension itself, which we understand independently of bodies and their properties, is the subject of mathematics. If we abstract from the bodies and their properties, the concept of infinitely extended, continuous, motionless, unchangeable and eternal space remains. This absolute space is, for Newton, the actual subject of pure mathematics. It encompasses all possible spatial areas that a body can occupy, i.e. materially occupy. How the ontology of mathematics is spelled out differs significantly between Descartes and Newton. However, Newton's absolute space, like the extension of bodies for Descartes, is part of physical reality. So, like Descartes, Newton advocates a non-Platonic realism. Newton's theory of knowledge in relation to mathematics resembles that of Descartes and can be understood as rationalist. According to Newton, we possess the ability to comprehend the actual and general nature of absolute space. This ability is not sensory for Newton, as our senses only allow us to perceive material bodies. Rather, this ability is an immediate clear mental vision. The statements of pure mathematics are therefore also a priori and not empirical for Newton. In our view, Newton's philosophy of mathematics shows that we should not see him as a pure empiricist.

We would like to offer a brief critical assessment of Newton's philosophy of mathematics here. We appreciate Newton's theory of knowledge of mathematics, as well as that of Descartes. However, we doubt that physical space is the subject of pure mathematics, and therefore cannot agree with Newton's ontology of mathematics. In our view, the nature of physical space cannot be determined without experiments and observation. Is physical space actually infinitely extended or is it limited? Is it truly continuous or in reality discrete with a granular fine structure? Is it flat or curved? All these questions cannot be answered with certainty without recourse to empiricism. We assume that the absolute space Newton has in mind is three-dimensional Euclidean space. Statements about this space are indeed a priori. We can either describe Euclidean space axiomatically or define it in a more general mathematical theory and then obtain the statements of Euclidean geometry by deduction. However, Euclidean space is not physical space, as all true statements about it can be justified independently of experience, which is not the case with physical space. We know from the experimental findings that support Einstein's general theory of relativity that three-dimensional Euclidean space is not even an optimal model of physical space. Even the best mathematical model of physical space that we have should not, in our opinion, be identified with it. Mathematical models of physical reality are subject to revisions based on new empirical findings.

4.4   Leibniz

Gottfried Wilhelm Leibniz (1646–1716) is sometimes referred to as the last great polymath, as his work includes contributions to philosophy, mathematics, physics, psychology, linguistics, jurisprudence, medicine and technology. Leibniz's writings are extensive, diverse and not systematically organised, a critical complete edition is still not finished.[9] Leibniz repeatedly revised his philosophical positions and his work contains contradictions, which he either did not notice or did not resolve due to lack of time.

A cornerstone of the metaphysics that Leibniz developed is the *Monadology (1714)*.[10] According to Leibniz, the world consists exclusively of monads, i.e. indivisible and unique units. Monads are the substances that make up the world. Space is constituted by the monads and does not exist, as in Newton, independently of the objects in it. Furthermore, space in Leibniz is not, as in Descartes and Newton, continuous and infinitely divisible. Rather, it is discrete and the monads are the smallest spatial units. Every spatial body is for Leibniz nothing other than a collection of monads that completely define the body. So far, we are reminded of contemporary elementary particle theories, which see the properties of physical space determined by the distribution of mass, or energy. If we follow Leibniz further, however, this impression shatters. Leibniz advocates panpsychism: monads are souls and capable of thinking. Lower monads only have dark, dreamy perceptions. Higher monads, such as the soul of a human, have consciousness and possess knowledge. The highest monad, namely God, has infinite consciousness and is omniscient. It is puzzling that Leibniz assumes that the mental processes of the monads more or less accurately reflect their external world, but arise entirely independently of the external world. In Leibniz's view, there is no causal connection between a monad and its environment. Instead, Leibniz posits a pre-established harmony between the perceptions of the monads and the world of all monads; this harmony enables knowledge and understanding. Given this metaphysical foundation, it is not surprising that Leibniz pays a lot of attention to supposed proofs of God's existence. Without the assumption of a powerful and benevolent creator God, the existence of the pre-established harmony is probably difficult to justify.

After this brief sketch of monadology, we now turn to mathematics and its philosophy in Leibniz's work. Like Newton, Leibniz develops the foundations of differential and integral calculus. The notation dy/dx for differential quotients and the integral symbol $\int$ that are still in use today are due to him. Leibniz published his results on differential and integral calculus three years before Newton in *Acta eruditorum (1684)*. However, Newton was already familiar with the basic features of infinitesimal calculus before Leibniz, and from today's perspective, both thinkers can equally be considered the founders of analysis. The convergence criterion for alternating series and a series representation of the Archimedean

---

[9] See Leibniz (2019) for the volumes published so far.

[10] See Leibniz (2017).

constant π are two results of Leibniz, which every mathematics student knows today.[11] In addition, the matrix notation in linear algebra, as well as a formula for calculating the determinant, can be traced back to Leibniz. Also, fundamental ideas of contemporary topology and fractal geometry are said to have been anticipated by Leibniz.[12]

In our view, Leibniz is not a realist in the philosophy of mathematics. Monads cannot be understood as objects of mathematics, since they are thinking substances. Although monads are indivisible, the identification of monads with points in a mathematical space is not in line with Leibniz's thinking. Also, the relations between monads are not the subject of mathematics. The pre-established harmony, which determines these relations, is metaphysical and not mathematical in nature. For Leibniz, the objects of mathematics, such as numbers and geometric figures, are rather clear and distinct ideas that originate from human imagination. They are therefore not independent of mental predecessors, but are constituted by mental processes of higher monads.

In the epistemology of mathematics, Leibniz advocates rationalism: mathematical truths are truths of reason. Leibniz characterises truths of reason as necessary truths, their negation leads to contradictions and they hold in all possible worlds. Truths of reason are especially a priori true, they cannot be refuted by experience. Leibniz contrasts truths of reason with truths of fact. Truths of fact can be confirmed or refuted by experience. According to Leibniz, there are two types of truths of reason, namely basic truths and derived truths. Basic truths need no justification because they are per se clear, and derived truths are obtained through logical deduction. The axioms of mathematical theory are basic truths for Leibniz, and the theorems of mathematical theories are derived truths.

Leibniz intended to develop a strictly symbolic language of the *characteristica universalis*, which should be capable of describing everything that is rationally ordered. A universal logic calculus, the *mathesis universalis*, should then allow us, to solve all problems that can be formulated in the *charteristica universalis*. This project, imbued with epistemological optimism, was unfortunately not realised by Leibniz. The ideas that Leibniz developed in this context find their echo in logicism and also in formalism in the philosophy of mathematics, which we will discuss in Chaps. 7 and 9.

To conclude this chapter, we feel that some critical remarks on Leibniz's philosophy of mathematics are appropriate. All non-realist approaches in the ontology of mathematics, which assume that the objects of mathematics are mental, have a common problem. They must explain why the mathematical perceptions and ideas of different people sufficiently coincide, to determine a unified subject area of mathematics. If such a unified subject area did not exist, a mentalistic ontology would imply that the statements of mathematics are purely subjective. Obviously,

---

[11] The beautiful Leibniz formula for π is: $\pi/4 = \sum_{k=0}^{\infty}(-1)^k/(2k+1) = 1 - 1/3 + 1/5 - 1/7 + 1/9 - \ldots$.

[12] See Mandelbrot (2014).

this claim is at least highly problematic, if not untenable. Leibniz provides an explanation for the agreement of the mathematical perceptions and ideas of the higher monads with his theory of pre-established harmony. However, we are not quite comfortable with this explanation. The existence of a pre-established harmony cannot be justified without strong theological assumptions.

We are quite willing to agree with Leibniz's thesis that true mathematical statements are truths of reason, and his distinction between basic truths and derived truths in mathematics is also sensible. Nevertheless, we have significant problems with the rationalist epistemology as described by Leibniz. We doubt that the mathematical basic truths, i.e., the axioms of mathematics, are necessarily true and hold in all possible worlds. The tautologies and inference rules of logic may indeed be necessarily true and therefore valid in all logically possible worlds. Whether mathematics as a whole can be reduced to logic, however, is questionable. We will discuss logicism in the philosophy of mathematics, for which Leibniz can be considered a precursor, in detail in Chap. 7. Here it should only be noted that from the assertion that a mathematical axiom is a priori true, i.e., can be justified independently of experience through immediate intuition, it does not follow that the axiom represents a logical truth. For example, a world in which there are only finite sets is possible. From the assumption that there are only such sets, no logical contradiction follows as far as we know. Nevertheless, we believe that the assertion of the existence of natural numbers is justified by immediate intuition, independent of experience.

Kantianism 5

Contents

5.1   Immanuel Kant and His Work

Immanuel Kant (1724–1804) is considered by some to be the most significant philosopher of modern times, his work is indeed extensive, diverse and original.[1]

Kant studied philosophy, physics and mathematics at the Albertus University in his hometown of Königsberg from 1740. After the death of his father in 1746, he worked as a private tutor and housemaster near Königsberg. In 1755, he qualified as a lecturer with the topic *The First Principles of Metaphysical Knowledge (Nova dilucidatio)* and became a private lecturer in Königsberg, with a very high teaching obligation (probably 16 SWS). Only in 1770 did he receive the call he had been striving for to the professorship for logic and metaphysics. From 1786–1788, Kant was rector of the Albertus University and became a member of the Berlin Academy of Sciences in 1787. In his last years, Kant came into conflict with the Prussian censorship authority. In an edict of 1794, he was accused of denigrating the Holy Scriptures and Christianity. Kant continued to teach until 1796, but was instructed to refrain from religious statements. Kant's life is usually described as uneventful, disciplined and uniform. He did not marry and died in 1804 without leaving any children.[2]

---

[1] We refer here to the complete edition Kant (1900–1908).

[2] More detailed biographical and historical information can be found, for example, in Störig (2016).

J. Neunhäuserer, *Introduction to the Philosophy of Mathematics*,
Mathematics Study Resources 22, https://doi.org/10.1007/978-3-662-72179-7_5

Already in his early writings, Kant turns away from the dogmatic metaphysics, which goes back to Gottfried Wilhelm Leibniz (1646–1716) and Christian Wolff (1679–1754).[3] In his habilitation of 1755, he develops an independent subjectivist theory of space and time, which forms a basis of his later transcendental idealism. By his own admission, it was the encounter with the sceptical philosophy of the Scot David Hume (1711–1776), which woke him from his dogmatic slumber. In his main work *Critique of Pure Reason (1781)* Kant develops a philosophy beyond rationalist dogmatism and empiricist scepticism.[4] In this work, we also find Kant's philosophy of mathematics, which we will discuss in detail in Sect. 5.2. As a preparation, we give here a brief introduction to the basic concepts and the structure of the Critique of Pure Reason.

Kant distinguishes between analytic and synthetic judgements. Analytic judgements are true or false solely on the basis of the meaning of the concepts contained in them. For example, the sentence *All bachelors are unmarried* expresses a true analytic judgement. Synthetic judgements are true or false on the basis of facts beyond language. True synthetic judgements thus express a knowledge about the world. Furthermore, Kant distinguishes a priori judgements from a posteriori judgements. True a posteriori judgements are based on sensory experience; they are empirical. In contrast, a priori judgements do not rely on experience; they are true or false independently of any experience. Apparently, all true analytic judgements are a priori true, they do not refer to the experiential world. Similarly, all a posteriori judgements are synthetic, they refer to the experiential world and therefore cannot be analytic. We summarise the types of judgements that Kant distinguishes in the following table:

	A priori	A posteriori
Analytic	Conceptual, logical	(None)
Synthetic	Scientific? Mathematical? metaphysical?	Empirical

The question underlying the Critique of Pure Reason is: How are synthetic a priori judgements possible? Kant is a rationalist insofar as he does not doubt the possibility of such judgements, but he is critical in the sense that he wants to clarify the conditions of the possibility of rational knowledge. The three concrete questions that the Critique of Pure Reason addresses are:

(1) How is pure mathematics possible?
(2) How is pure natural science possible?
(3) How is metaphysics as a science possible?

---

[3] See also the last Chap. 4.
[4] We refer here again to Kant (1900–1908).

These questions are particularly addressed in the transcendental doctrine of elements, which forms the first part of the Critique of Pure Reason. The transcendental doctrine of method in the second part contains the sketch or the necessary conditions for a complete philosophical system based on the transcendental doctrine of elements. The transcendental doctrine of elements consists of three long sections, the transcendental aesthetic, the transcendental analytic and the transcendental dialectic, with the last two sections being summarised under transcendental logic. In the transcendental aesthetics, Kant attempts to clarify the basis of the possibility of pure mathematics; this part will occupy us in Sect. 5.2. In the transcendental analytics, Kant attempts to justify the possibility of pure non-empirical natural science. The transcendental dialectic aims to clarify the way in which metaphysics is possible as a science. For Kant's philosophy of mathematics, these two parts are of secondary interest and we will therefore not consider them in the next section.

The impact of the Critique of Pure Reason is remarkable. It is the decisive starting point of German Idealism, whose most important representatives are Johann Gottlieb Fichte (1762–1814), Georg Wilhelm Friedrich Hegel (1770–1831) and Joseph Schelling (1775–1854). The Critique of Pure Reason is also the central reference point for Neo-Kantianism in the nineteenth century. Meanwhile, the Critique of Pure Reason is globally received. On the other hand, theists who are stuck in traditional metaphysics attack the work. Kant's systematic critique of this metaphysics in the transcendental dialectic shows, among other things, that the existence of an immortal soul or God as the primal ground of the world cannot be proven by the means of pure reason. The German philosopher Moses Mendelssohn (1729–1786) therefore refers to Kant as the "all-crushing". For some Christians, Kant's philosophy was probably unacceptable and the Critique of Pure Reason even made it onto the Index librorum prohibitorum in 1827, a list of books whose reading was a grave sin for every Catholic.[5]

After the Critique of Pure Reason, Kant published a series of further works. In the *Groundwork of the Metaphysics of Morals (1785)* and the *Critique of Practical Reason (1788)*, Kant develops his ethics. The famous categorical imperative is introduced in the first book and extensively justified in the second book. The *Critique of Judgement (1790)*, which attempts to mediate between the two previous critiques and surprisingly offers a philosophy of feeling, concludes the critical writings. Finally, Kant's late work *Perpetual Peace (1795)* should be mentioned, in which he applies his moral philosophy to politics and develops the idea of a federation of nations that secures global peace.[6] This idea is now found in the Charter of the United Nations.[7]

Whatever one's individual stance on Kant's philosophy may be, his work is undoubtedly influential and impressive.

---

[5] See Fischer (2005). The Index was a tool of the Roman Catholic Inquisition, which was not abolished until the Second Vatican Council in 1966.

[6] For the works, we again refer to Kant (1900–1908).

[7] See Pollok (1996).

5.2 Mathematics in the Critique of Pure Reason

As stated in Sect. 5.1, Kant develops his philosophy of mathematics in the transcendental aesthetics, the first part of the Critique of Pure Reason. By aesthetics, he does not mean the study of beauty. Kant is rather referring to the Greek term *aisthesis*, i.e., sensory perception. The transcendental aesthetics attempts to clarify the conditions of the possibility of sensory perception. For Kant, our sensibility consists in the ability to receive representations through the senses. Kant calls such representations intuitions; they are given to us directly through the affection of the senses. Now the question arises as to what mathematics has to do with our intuitions, which are given through the senses. Kant holds the following original and strange ontological and epistemological position in the philosophy of mathematics:

(1) The objects of mathematics are a priori forms of our intuition.
(2) These forms are the basis of mathematical knowledge.

To understand these theses, we need to explain which a priori forms Kant attributes to our intuitions. Kant distinguishes two types of sensibility and thus two forms of intuition. On the one hand, we have the five external senses of seeing, hearing, smelling, tasting, touching. The form of the representations given by these senses, is space for Kant. On the other hand, we have an internal sense, introspection, which gives us our own mental states or processes. The form of the representations given to us by the internal sense is time. Time is also a form of the representations of the external senses, as these representations are also given to us introspectively. Taken together, space and time are thus the two forms of intuition.

In mathematics, Kant now also knows two areas, namely geometry and arithmetic. As his examples show, by geometry he always means the elementary geometry of Euclid (third century BC), which he developed in the *Elements*.[8] By arithmetic, Kant always means the elementary arithmetic of natural numbers, where he holds the view that the natural numbers are given by the process of counting. Analysis, algebra, probability theory and numerical analysis, which were also already present in Kant's time in their initial stages, are not considered in the Critique of Pure Reason. Kant's idea in the philosophy of mathematics now consists in a simple assignment between areas of mathematics and forms of sensibility. Geometry deals with space, more precisely with the spatial form of our intuitions, i.e., the intuition space. The basis of arithmetic for Kant is time, more precisely the temporal form of our intuitions, i.e., the intuition time. The succession of moments allows counting, and counting is supposed to constitute the natural numbers, which are the subject of arithmetic.

Starting from this approach, a philosophy of space and time also represents a philosophy of mathematics, and the philosophy of space and time forms the core of transcendental aesthetics. Kant tries to show that space and time are necessary and

---

[8] See Mollweide and Lorenz (2017).

universal. The objects that our senses give us are extended, and they can only be so if our perception has spatial form. Space is therefore a condition of the possibility of sensory experience and in this sense necessary. Furthermore, the spatial form applies to every perception of objects and is therefore universal. We can only experience in a temporal sequence of perceptions. Time is therefore also a condition of the possibility of experience and necessary. Finally, every perception is temporally ordered and time is therefore universal. Now Kant argues as follows: What is given to us empirically through the senses cannot be necessary and universal. But space and time are necessary and universal. Therefore, space and time are given to us a priori, independent of sensory experience and as a necessary prerequisite for this. Space and time precede all experience in the sensory faculty of our reason and order any empirical material. Judgements about objects that are given to us a priori are a priori judgements, they do not require justification through experience. True judgements about space and time, in Kant's sense, are therefore a priori true, their justification lies in the a priori form of our perceptions. Since mathematics, according to Kant, deals with the space of perception and the time of perception, the judgements of mathematics are a priori and not empirical. Kant also tries to show that judgements about space and time and thus the judgements of mathematics are not analytical, but synthetic. An analysis of the concepts of space and time does not enlighten us about the properties of space and time, i.e. the forms, our perception. Space and time are ultimately not purely conceptual, but intuitive. Judgements about space and time cannot therefore be justified by a concept analysis alone, they are not analytical. Such judgements tell us something about the world, more precisely about the forms in which the world necessarily appears to us. As a whole, Kant believes he has shown that the judgements about space and time and thus the judgements of mathematics are synthetically and a priori justified. Their source are the forms of our sensibility, which structure our experiences.

The interpretation of the transcendental aesthetics, which we offer here, is certainly incomplete and somewhat simplified. We would be able to elaborate and exegetically justify this more precisely, but do not believe that this is necessary and appropriate here. Since the style of Kant's explanations is sometimes somewhat cryptic, interpretations of Kant are notoriously controversial. Einstein once said that everyone has their own Kant.[9] Nevertheless, we are confident that we have correctly outlined the essence of Kant's philosophy of mathematics here.

5.3   Critical Assessment

In contemporary philosophy, the sharpness of both the distinction between analytical and synthetic statements as well as the distinction between a priori and empirical truth or justification is often questioned.[10] We are not able to understand

---

[9] See Rosenthal-Schneider (1988).

[10] See Chap. 12 for this.

this, as these distinctions can be explained clearly and understandably. Like Kant, we believe that there are synthetic statements that are a priori true and do not require justification through sensory experience. We also agree with Kant that it is a central task of philosophy to determine which statements these are. However, we doubt that there are synthetic statements of natural sciences that are a priori justified. A statement of natural sciences refers to nature and this is apparently only accessible to us empirically, by referring to sensory experiences. In this respect, our epistemological attitude is clearly empiricist. For the assessment of Kant's philosophy of mathematics, however, this is not relevant. The interesting question is whether or which statements of mathematics are not analytical and how these statements can be justified a priori. Logizism, which we discuss in Chap. 7, tries to reduce all of mathematics to logical-analytical statements. We will see that part of arithmetic may actually be reconstructed purely analytically. With regard to all of mathematics, however, logizism failed; there are mathematical statements that are certainly not analytical. If we assume that statements of mathematics cannot be justified by sensory experience, there are thus statements of mathematics that are synthetic and a priori. In our opinion, these statements even make up the majority of mathematics, we will go into more detail on this in the next chapter. We largely agree with the credo of Kant's philosophy of mathematics, that the statements of mathematics are synthetic and a priori. Nevertheless, we regret to note that Kant's ontology and epistemology of mathematics are misguided. Mathematics does not deal with the forms of our perception and these forms do not justify any mathematical statements or theories. This is quite easy to see. Mathematics investigates a variety of spaces with different dimensions, geometries, and structures.[11] Which of these spaces accurately describes the spatial form of our perception is not a mathematical question. Mathematics does not, in fact, concern itself with this question and has no methods to provide an answer. It could be the task of perceptual psychology to determine which mathematical model best describes the spatial form of our perception. Even if a theory of perception can identify a specific mathematical space as the spatial form of our perception, this is irrelevant to mathematics. Such a finding does not justify a mathematical theory. It is not even a good reason to prefer the study of a particular mathematical space over the study of other spaces. The same applies to the temporal form of our perception. Mathematics provides many possible models that can be used to describe the temporal form or shape of our perceptions. The arithmetic of the temporal form of our perception could be continuous or discrete, it could be unlimited or cyclic. Perceptual psychology could consider many mathematical models of the temporal form of our perception. What a good description of this form looks like is not a topic of mathematics and results in this regard are irrelevant to mathematical research and cannot be used to justify mathematical statements. Overall, the space-time form of our perception is no more a subject of mathematics than physical space-time. The

---

[11] For example, see the terms defined in Neunhäuserer (2017).

former is a subject of psychology, the latter is a subject of physics. The results of psychology and physics are unsuitable as justification for mathematical theories.

Perhaps Kant's philosophy of mathematics is more understandable from a historical perspective. In his time, non-Euclidean geometries, topological spaces, high- and infinitely dimensional vector spaces, etc. were not yet known. Also, cyclic arithmetic or modular arithmetic and the theory of real numbers, i.e., continuous arithmetic, were still in their infancy. Some simple considerations, however, already show that the shape of space and time, as forms of our perception, are not necessarily and universally, and thus not a priori, given.

If we reduce sensibility to taste and smell, the form of representations given by these senses is probably in no way spatial. Nevertheless, experiences are possible solely through these senses, and even objects can be identified and recognised as taste and smell entities in a temporal succession. Spatiality is not a necessary condition of sensory experience. The form of perceptions given by sight is spatial. This statement is analytic, as the concept of seeing implies some kind of spatiality. However, the closer determination of the type, structure or shape of this spatiality is not a priori possible; no specific geometry is a necessary condition of seeing. Here are two simple examples that would have been accessible to Kant: The spatial form of our perceptions is three-dimensional. Kant was already familiar with a two-dimensional geometry, namely the geometry of the Euclidean plane. The existence of a subject that only sees two-dimensionally, whose perceptions therefore have a spatial form that can be described by a Euclidean plane, is possible and cannot be ruled out a priori. Subjects who see in four or more dimensions are also conceivable; the corresponding geometries were not yet known to Kant. The three-dimensionality of the spatial form of perceptions is not a condition of the possibility of experience, it is an empirical fact. Our second example shows that properties of the spatial form of our perceptions can even change. In every plane of Euclidean geometry, there is exactly one line in the plane that goes through the point and does not intersect the line. Euclidean geometry seems to describe the spatial form of our perception very well, but in some cases it does not. If we limit our perception to a room and consider the floor of the room. For a line and a point on the floor, there seem to be infinitely many lines that go through the point and do not intersect the line, after all, lines in our perception end at the walls of the room, see Fig. 5.1. Mathematically formulated, the spatial form of our perception is sometimes compact and sometimes open. Consider the form of our perception in a place where we can see the horizon in all directions, for example on the top of a mountain. Any two lines in the horizontal plane seem to have an intersection point, with the intersection point of some lines being found on the horizon.[12] In both cases, Euclidean geometry does not seem to be a very good model of the form of our perception. Be that as it may. The form of

---

[12] Euclid seems to be aware of this problem of his geometry and speaks of parallels intersecting at an infinitely distant point, see Mollweide and Lorenz (2017). From today's perspective, there is either an intersection point between two lines or there is none, and Euclidean geometry is the one in which there are clearly defined parallels that do not intersect.

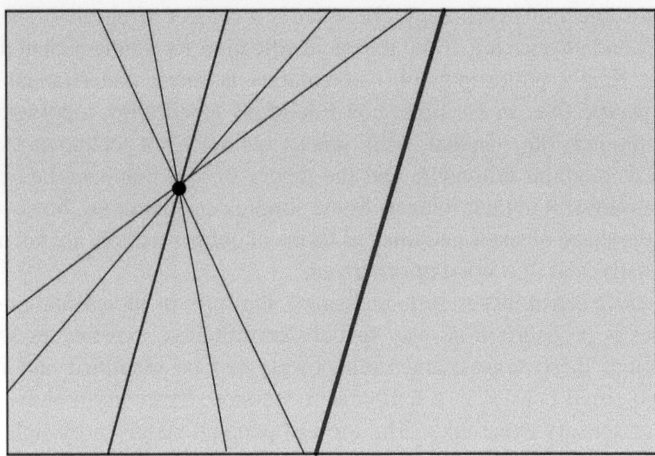

Fig. 5.1 Geometric shape of a limited perception

our perception is by no means necessarily Euclidean; and when Euclidean geometry is a good model is an empirical question that cannot be answered independently of experience.

That our perceptions have some temporal form may indeed be a condition of the possibility of experience in the sense of change of perceptions. This is an analytic thesis that can be derived from the meaning of the concept of change. More importantly, the shape or structure of time is not given to us a priori. A sequence of time points as the temporal form of our perception is by no means the only option; a continuous temporal form of our perception is just as experiential. We assume that it is a psychological question under which conditions a discrete and under which conditions a continuous model of perceptual time is more appropriate. If the subject is subject to an external clocking, a discrete model may be appropriate. However, if we are on holiday lying on a sun lounger and looking at the sea, a continuous model is certainly more suitable. Kant probably did not notice this, as his lifestyle was very disciplined. Also, a cyclic shape of the temporal form of our perceptions becomes experiential when our perceptions are determined by cyclical natural phenomena such as day and night or the seasons.

Simple considerations thus show that space and time as forms of our perception are not immutable and not given a priori. If we adhere to a rationalist theory of knowledge of mathematics, perceptual space and perceptual time are not the subject of mathematics. We see Kant failing as a philosopher of mathematics. One cannot simultaneously claim that the statements of mathematics are synthetic a priori, and refer to the forms of our perception.

Mathematics in German Idealism 6

Contents

6.1   Introduction

German Idealism is a philosophical movement that developed around the turn of the eighteenth to the nineteenth century. The main representatives of German Idealism are Johann Gottlieb Fichte (1762–1814), Georg Wilhelm Friedrich Hegel (1770–1831) and Friedrich Wilhelm Joseph Schelling (1775–1854). Some authors also count Immanuel Kant (1724–1804) among the German Idealists, although a systematic distinction is offered here. The common feature of the German Idealists is the assumption that mental entities are ontologically fundamental. There are neither concrete physical nor abstract ideal objects that exist beyond and independently of the mind. Kant can be understood as a precursor of German Idealism, although he assumes the existence of a *thing in itself* that is not part of the mental world. This assumption is indispensable for his conception of causality and is rejected by the Idealists. We have already discussed Kant's philosophy of mathematics in detail in the last chapter.

Fichte's Idealism starts from an *absolute ego* that sets its own being in an act of creative doing itself. The ego also generates the non-ego, i.e. the objects of the world, within itself. This is an unconscious, free and groundless process, motivated by the infinite creative power of the ego, which can only be realised through a contrast. The objects of mathematics, such as numbers and geometric figures, would accordingly be a product of the creative ego. A similar thought will encounter us

© The Author(s), under exclusive license to Springer-Verlag GmbH,   47
DE, part of Springer Nature 2025
J. Neunhäuserer, *Introduction to the Philosophy of Mathematics*,
Mathematics Study Resources 22, https://doi.org/10.1007/978-3-662-72179-7_6

again in Chap. 8 in the philosophy of mathematics by Luitzen Brouwer (1881–1966). However, Fichte did not develop a philosophy of mathematics and does not seem to have been particularly interested in mathematics or natural sciences. In contrast, Schelling and Hegel explicitly refer to the mathematics and natural sciences of their time.[1] Therefore, in the next two sections of this chapter, we will focus more on Schelling and Hegel, but not further on Fichte.

German Idealism in philosophy is closely linked to German Romanticism, an influential artistic-intellectual movement in the late eighteenth and early nineteenth century. As is well known, the Romantics prefer feeling to mathematical thought. The poets, painters and musicians of Romanticism therefore usually encounter mathematics with disinterest or even rejection.[2] An exception is the writer and cultural philosopher Karl Wilhelm Friedrich Schlegel (1772–1829), to whom we dedicate the last section of this chapter.

6.2   Schelling

The philosopher Friedrich Wilhelm Joseph Schelling was born in 1775 in Leonberg in Württemberg and studied from 1790 together with Friedrich Hölderlin and Georg Wilhelm Friedrich Hegel theology in Tübingen and later mathematics and natural sciences in Leipzig. From 1798 to 1803, Schelling taught alongside Fichte and Hegel at the University of Jena. During this time, his main philosophical works *First Draft of a System of Natural Philosophy (1798/99)*, *System of Transcendental Idealism (1800)* and *Presentation of my System of Philosophy (1801)* were published.[3] Subsequently, Schelling increasingly turned away from the sciences and philosophy, in the sense of a rational reference to the world. He worked as an academic in Würzburg, Munich and Erlangen and succeeded Hegel in Berlin in 1841. In 1842 and 1845 he gave the lecture *Philosophy of Mythology and Revelation*. Schelling's later works are characterised by a return to Christian faith and a mystical-religious worldview. Schelling died in 1854 during a spa stay in Bad Ragaz in Switzerland.

At its core, Schelling's Idealism is a philosophy of identity. Among other things, he asserts the identity of subject and object, of mind and nature, as well as the identity of thinking and being. The absolute identity of subject and object in Schelling's philosophy is based on the view that a subject cannot exist without an object and an object cannot exist without a subject. The unity of subject and object now consists in the fact that they mutually bring each other into existence. One form of this unity is manifested in the self-consciousness of the human subject, which precedes the identity of subject and object as an absolute.

For Schelling, nature is not the material reality of particles and their interactions, rather he believes he can discover everywhere in nature the unconscious activity of

---

[1] We refer to Störig (2016) for an introduction to German Idealism and Fichte's philosophy.

[2] See for example Novalis' famous poem "When no more numbers and figures".

[3] We refer here to the complete edition Schelling (1856–1861).

the mind. The mind now does not act as an external force on nature, but it is itself nothing other than the becoming self-conscious of nature. On a fundamental level, nature and mind are therefore supposed to be identical. Schelling's philosophy is popular among the German Romantics, as they want to overcome contrasts and are inspired by the search for unity. To this day, the *oneness with nature* is a popular romantic motif.

The identity of thinking and being is fundamental to Schelling's philosophy of mathematics. He explicitly asserts that mathematical thinking and being are one. The unity of thinking and being in geometry is supposed to be approached from the side of being and in arithmetic, starting from the side of thinking. Schelling believes that he has explained the difference and the commonality of these mathematical disciplines with this.[4] The distinction thus obtained between arithmetic and geometry is difficult to understand and is hardly likely to convince today.

In the following, we want to show that Schelling's ontological identification of mathematical being with mathematical thinking brings considerable problems with it. Firstly, not every mathematical thought can be identified with a mathematical fact or a mathematical truth. Mathematicians make mistakes, they sometimes consider statements to be true, even proven, which are false and thus do not describe any part of mathematical being. To accommodate Schelling at this point, we could assume that he wants to assert the identity of mathematical knowledge and mathematical being. To understand this claim, we would need a theory of knowledge of mathematics that specifies what distinguishes mathematical knowledge, i.e., the content of true mathematical statements. Unfortunately, we cannot find a theory of knowledge of mathematics in Schelling's work. The absolute identity of the knowing subject and the known object, which Schelling asserts, contradicts any theory of knowledge. It has the bizarre consequence, in particular, that with the disappearance of all knowing subjects, all objects of knowledge also disappear, and the world thus dissolves into nothing. Even if we disregard Schelling's bizarre metaphysics, the identity of mathematical knowledge and mathematical being can only be reconciled with a theory of knowledge of mathematics at a high price. A theory of knowledge of mathematics cannot ignore that we gain mathematical knowledge through mathematical research. If mathematical knowledge and mathematical being are one, it follows that mathematical research has ontological consequences and mathematical objects or facts are not discovered, but put into the world. This consequence is in line with the philosophy of Fichte, which we briefly discussed in the introduction. The epistemological problems of this attitude in the philosophy of mathematics will occupy us further in Chap. 8 on intuitionism and also in Sect. 13.3 on fictionalism. At this point, we only want to point out that scientific research, in the usual sense of the term, does not itself produce the objects it investigates.

---

[4] See Works I.4, p. 347 in Schelling (1856–1861).

6.3   Hegel

Georg Wilhelm Friedrich Hegel was born in Stuttgart in 1770 and studied with Schelling and Hölderlin in Tübingen. From 1801 to 1807 he taught at the University of Jena and there wrote his first significant philosophical work, the *Phenomenology of Spirit (1807)*. As rector of a grammar school in Nuremberg, he completed his second major work, the *Science of Logic (1812/13/16)*, in three volumes. In 1818, Hegel was called to a professorship in Berlin. There he dealt, among other things, with the *Elements of the Philosophy of Right (1820)* and advanced to become the Prussian state philosopher. Far beyond Hegel's death in Berlin in 1831, the Hegelian school dominated German philosophy. Among other things, Hegel's philosophy had a decisive influence on Marxism.

Hegel's work is complex, powerful in words and is considered by some to be largely incomprehensible.[5] Hegel's idealism is characterised by the basic assumption that the entire world process is the self-unfolding of the spirit and the philosopher studies this self-unfolding. First, the world spirit is in the state of *being-in-itself*, then it alienates itself in the form of spatio-temporal nature and enters the state of *being-other*. In the third and final stage, the spirit returns to itself and enters the state of *being-in-and-for-itself*. The philosophical discipline that studies the first state is, according to Hegel, logic, the philosophy of nature studies the second state of the unfolding of the spirit and the philosophy of the spirit has the spirit returned to itself as its object.

For Hegel's philosophy of mathematics, his logic, i.e., the *being-in-itself* of the spirit, is decisive. Mathematics is for Hegel a pure science of understanding, which is not dependent on observations of nature. In arithmetic and even in geometry, for Hegel, inferences from assumptions and not forms of our intuition are fundamental.[6] He clearly distinguishes himself from Kant's philosophy of mathematics, which we presented in the last chapter. A conceptual philosophical foundation of mathematics seems to Hegel to be particularly necessary in relation to higher mathematics. This refers to differential and integral calculus, which was still in development in Hegel's time and operated with infinitely large and infinitely small quantities.

Hegel believes he can give mathematics an epistemological foundation with his logic. One might now be tempted to read and appreciate Hegel as a precursor of the logistic philosophy of mathematics, which we will discuss in the next chapter. If one looks a little more closely at Hegel's dialectical logic, this reading unfortunately becomes invalid. Hegel assumes that a thesis can be true just like its antithesis and that this contradiction is resolved in a synthesis, with which both the thesis and the antithesis have unrestricted validity. Hegel expressly rejects the principle of non-contradiction, which states that a statement and its negation cannot both

---

[5] We refer here to the complete edition Hegel (1970). The German philosopher Arthur Schopenhauer (1788–1860) said of Hegel's work: "However, the greatest audacity in serving up bare nonsense, in smearing together meaningless, raging word tangles, as had hitherto only been heard in madhouses, finally appeared in Hegel...."

[6] See IX, pp. 46–47 in Hegel (1970).

be true at the same time, in his logic. Most theorists of science and almost all scientists are of the view that a logic that denies the law of non-contradiction is not suitable as the basis of a science.[7] We emphatically agree with this view. Theories that contain contradictions do not create knowledge but confusion. It is a fundamental scientific principle that such theories must be revised and replaced by non-contradictory theories. Anyone who denies the law of non-contradiction, bids farewell to science and philosophy, in the sense of a rational reference to the world, and can only succeed in the fine arts, religion or politics. For mathematics, the law of non-contradiction is of particular importance. A propositional logic or predicate logic, which mathematicians have been using implicitly or explicitly in their proofs since Euclid and to this day, is unthinkable without it.[8] Therefore, Hegel makes no relevant contribution to the epistemology of mathematics. Nevertheless, we still want to discuss the ontology of mathematics in his *doctrine of being* from the *science of logic*.

The fundamental object of mathematics for Hegel is a non-quantitative unity, which he calls *the one*. This is an abstract object, which is distinguished only by its identity with itself. From such units, multiplicity arises through the abstract relations of repulsion and attraction. Numbers are now determined by the number of units in a multiplicity, which itself again forms a unity. We are reminded here of a modern set theory, which demands the existence of an empty set as a fundamental object and constructs the natural numbers from it.[9] It is therefore tempting to attribute a Platonic ontology of mathematics, or at least of arithmetic, to Hegel. Fundamental objects of mathematics are abstract entities that exist independently of physical and mental processes. This interpretation stands or falls with whether, for Hegel, being or the *being-in-itself* of the world spirit, exists independently of the development towards *otherness* and *being-in-and-for-itself*. Does the being of mathematical objects for Hegel exist independently of becoming? Hegel's position on this question seems unclear to us. Probably his dialectical logic prevents a clear statement. The thesis *The objects of mathematics are unchangeable and timeless* is opposed by the antithesis *The objects of mathematics are changeable and temporal*. Hegel tries, as said, to resolve such contradictions through a synthesis, which is logically not possible.

In his *doctrine of being*, Hegel deals extensively with the concept of infinity. He rightly states that it is a qualitative and not a quantitative concept. Mathematics therefore deals not only with quantitative but also with qualitative properties of its objects. From today's perspective, Hegel's statement may seem trivial. Whether a set is finite, countably infinite or uncountably infinite is a qualitative property of the set. Hegel's insight is remarkable at this point, the precision of the concept of infinity was only achieved by Georg Cantor (1845–1918) towards the end of the nineteenth century. Although school mathematics sometimes gives a different impression,

---

[7] A prominent critic of dialectical logic is Popper (1940).

[8] We provide an introduction to logic in Sect. 7.3.

[9] We are guided here by the modern reading of Hegel by Pinkard (1981).

modern mathematics actually deals predominantly with qualities, such as with the algebraic properties of sets with operations, the completeness or compactness of spaces or the structural properties of operators, etc. In this respect, Hegel was on the right track. It is now curious to note that the Marxist philosophy of mathematics refers to Hegel and attributes a materialistic ontology of mathematics to him, precisely because he takes into account qualitative aspects of mathematics.[10] Such an argument assumes that qualitative properties are more likely to belong to material than to abstract or mental entities, which does not seem to be the case. The Marxist interpretation of Hegel is probably determined by their materialistic ideology and untenable. However, we could imagine that Hegel would have agreed with the statement: *The objects of mathematics are abstract in their being-in-itself and material in their otherness*. This sentence may sound good, but it does not tell us what kind of objects mathematics investigates.

6.4   Schlegel

Karl Wilhelm Friedrich Schlegel was born in Hanover in 1772. He studied in Göttingen, Leipzig and from 1795 in Jena, where he qualified as a lecturer in 1800 and gave lectures on transcendental philosophy, which Hegel is said to have attended. During his time in Jena, Friedrich Schlegel, together with his brother August Wilhelm Schlegel (1767–1845), Novalis (1772–1801), Ludwig Tieck (1773–1853) and others, founded German Romanticism in literature. During this time, his only novel, *Lucinde (1799)*, was created. After the disintegration of the Jena Romantic circle, Schlegel lived in Paris from 1802 to 1804 and after his marriage from 1804 to 1808 in Cologne. During this time he devoted himself intensively to Indology and became with the work *On the Language and Wisdom of the Indians (1808)* a pioneer of comparative linguistics. In 1809, Schlegel became secretary of the court and state chancellery in Vienna and advanced as editor of *Concordia* to a central figure of the Catholic-influenced late Romanticism in Vienna. Schlegel's later work deals with the *Philosophy of Life (1828)*, the *Philosophy of History (1829)* and is mythically-religious in nature. Schlegel died in 1829 of a severe stroke on a trip to Dresden, where he was to give a lecture on his *Philosophy of Language (1829)*.

Schlegel did not leave behind a systematic philosophy like Fichte, Schelling and Hegel, his philosophical considerations are scattered in essays, lectures and fragments.[11] A fundamental concern runs through his entire work. Schlegel wants to bring together science and art through an interdisciplinary approach and unite them into a higher whole. This search for unity and the attempt to resolve contrasts and contradictions is a characteristic motif of German Romanticism. The philosophical foundation of this concern can be found in idealistic philosophy.

---

[10] Kolman and Yanovskaya (1931) is the strangest text on the philosophy of mathematics we have read.

[11] We refer here to the complete edition Schlegel (1958ff).

The art of all arts, thus the most general art for Schlegel is poetry, which is not only found in literature, but also in music. Schlegel's poetics now draws on mathematics. The poetic unity as the identity of the different is represented by mathematical equations or functions. The poetic ideal is to be determined by an infinite approximation, thus a kind of limit.[12] Mathematical formulas, thus statements of mathematics, have a poetic character for Schlegel, they become a part of art. Schlegel deliberately overlooks the fact that the sciences and especially mathematics are not free like art and poetry. Poetic sentences do not require a justification founded in epistemology, a large part of mathematics, however, consists of the struggle to prove theorems. Similarly, a large part of empirical sciences consists of collecting data, with the aim of supporting hypotheses. We expect mathematical or general scientific sentences to express facts, we do not expect this from poetic sentences. So what should the unity of science and art look like? Schlegel gives us an answer to this question. The science that unites all arts and sciences into one and thus represents the union of science and art is magic. Since for Schlegel the principles of mathematics are magical, he goes so far as to identify mathematics and magic. It is disturbing, but the sentence *Mathematics=Magic* can be found in Schlegel's work.[13] It is difficult to take this position seriously in the philosophy of mathematics. The author of this book does not know that he ever applied magical practices as a mathematician and he is also not aware of any colleagues who practice magic. Nevertheless, we would like to explore some aspects of mathematics here, which may seem magical. If a proof of a mathematical theorem succeeds in an unexpected way or if a surprising connection between subfields of mathematics appears, this can seem like a miracle or indeed like magic. This phenomenon is more of psychological than of philosophical relevance and easy to explain. We only have a rudimentary understanding of the network of logical implications of our mathematical axioms, definitions and theorems. Therefore, we are always surprised in mathematical research. If a logical connection or an original proof establishes itself over time, the impression of dealing with magic is relativised and usually disappears. Another remarkable aspect of mathematics is of greater philosophical relevance. We seem to be able to understand certain fundamental mathematical facts directly, without needing a logical justification. This phenomenon is known as mathematical intuition. We will see in the next chapter that mathematics as a whole cannot be reduced to logic. Mathematical intuition thus seems to be indispensable in the foundations of mathematics. This aspect of mathematics hardly fits into a purely naturalistic or physicalistic worldview and will surprise and challenge proponents of such a worldview. We will return to this topic in Chap. 12. However, we do not believe that mathematical intuition is a miracle and has something to do with magic, and instead suggest understanding it as an essential and irreducible part of our rationality. The mystery of mathematical intuition remains a phenomenon that Schlegel could have referred to, but to our knowledge he did not.

---

[12] See Schlegel (1958ff) 16, p. 148.

[13] See Schlegel (1958ff) 19, p. 10.

Logicism

7

Contents

7.1   The Logicist Position

The logicist position in the philosophy of mathematics, in short, assumes that mathematics can be reduced to a sufficiently comprehensive formal logic and therefore mathematics is a part of logic. The basic assumptions of logicism can be formulated as follows:

1. The concepts of mathematics can be defined using the concepts of a suitable formal logic.
2. The sentences of mathematics follow from purely logically evident premises by means of logical deduction.

In other words, the first assumption of the logicist philosophy of mathematics states that the vocabulary of mathematics is a subset of the vocabulary of logic, and the second assumption states in this sense that the set of mathematical theorems is a subset of the set of theorems of logic. Logicism is often not represented in relation to all of mathematics, but only in relation to a subfield of mathematics. Logicism in relation to a subfield of mathematics asserts that the concepts of this subfield can be traced back to concepts of logic and that the sentences of the subfield can be derived from logically evident premises, the axioms of logic. We will see below

J. Neunhäuserer, *Introduction to the Philosophy of Mathematics*,
Mathematics Study Resources 22, https://doi.org/10.1007/978-3-662-72179-7_7

that in particular logicism in relation to arithmetic plays a decisive role in the philosophy of mathematics. That in mathematics theorems are proven from axioms by means of logical deduction is rarely disputed. A mathematical proof is nothing other than a derivation of a theorem using the tools of logic from explicitly given assumptions. However, the claim of logicism is considerably stronger. The axioms of a mathematical theory themselves should be derivable from suitable definitions and tautologies, i.e. purely logically evident premises.

At first glance, the fundamental objects of mathematics, such as numbers, sets, relations and functions, Logic and mathematical concepts are not the subject of logic and are not defined in the terminology of logic. Similarly, it is not obvious that the axioms of mathematical theories are propositions that are evident for logical reasons, or can be derived from tautologies. A logicist in relation to a subfield of mathematics must therefore undertake the task of reducing the mathematical theory to logic. This means providing bridging definitions that translate the vocabulary of the mathematical theory to be reduced into the vocabulary of the logical theory, and in the next step showing that the axioms of the mathematical theory follow from the axioms of the logic used.

A successful reduction of a field of mathematics to logic represents a major epistemological advance. The true statements of logic are, to use Kant's terminology, analytically and a priori true, see also Chap. 5. Such statements are true solely on the basis of the meaning of the linguistic expressions used and independent of any experience. The degree of certainty and trustworthiness of statements that are actually analytically and a priori true is hard to surpass. After a successful reduction of a subfield of mathematics to logic, the statements of this subfield of mathematics would have the same degree of certainty and trustworthiness, as these statements proved to be analytically a priori. As we will see in Sect. 7.2, it was a strong motivation for mathematicians and philosophers who pursued the logicist project of reducing mathematics to logic to gain certainty in the foundations of mathematics. On the other hand, analytically true statements convey no knowledge about the world beyond the logical relationships of the terms used. If logicism in the philosophy of mathematics were true, we would learn nothing from mathematical sentences that goes beyond our mathematical terminology and its logical connections. Some researching mathematicians may feel that they are making a discovery about a world that goes beyond the logical relationships of mathematical concepts with the proof of a new mathematical theorem. If logicism were true, this feeling would prove to be unfounded. However, in this chapter we will show that even the reduction of arithmetic to logic is extremely problematic, we seem to need certain premises that are either not tautologies or whose truth is questionable. A reduction of all mathematics to logic seems impossible today.

7.2   **Historical Development**

The German philosopher and mathematician Gottfried Wilhelm Leibniz (1646–1716) was probably the first to see the ideas and principles of logic as the basis of any science. Leibniz formulated a symbolic logic and represented the traditional conceptual logic of Aristotle with it.[1] Logicism in the philosophy of mathematics essentially goes back to the German mathematician Richard Dedekind (1831–1916). Dedekind's concern was to give mathematics in general and analysis in particular a secure foundation in logic. This foundation should be independent of our mathematical intuition and our perceptions of space and time.[2] In this sense, Dedekind's thinking turns against both Platonism and Kantianism in the philosophy of mathematics. Dedekind succeeded in constructing the real numbers, the basis of analysis, from sets of rational numbers and providing an axiom system that characterises the natural numbers and thus forms the basis of arithmetic.[3] The axiom system of natural numbers by Dedekind is equivalent to the Peano axioms used today by the Italian mathematician Giuseppe Peano (1858–1932).[4] In terms of logicism, however, it should be noted that Dedekind does not ground either the concept of set he uses or the axiom system of natural numbers in logic.

The main representative of logicism in the nineteenth century is the German logician and philosopher Gottlob Frege (1848–1925). In his main works *Foundations of Arithmetic (1884)* and *Basic Laws of Arithmetic (1893)* he derived the axioms of natural numbers from the formal logic he developed.[5] To provide a foundation for arithmetic in a suitable logic and to formalise this very logic seems to have been Frege's life theme. At the beginning of the twentieth century, however, the British mathematician and philosopher Bertrand Russell (1872–1970) noticed that a principle of Frege's about sets as the scope of a concept is untenable, as it leads to paradoxes.[6] We will go into this in detail in Sect. 7.4. It is reported that Frege was deeply affected by Russell's discovery and abandoned the logicist project. We see Gottlob Frege almost as a tragic figure in the philosophy of mathematics. Bertrand Russell developed together with the British Mathematician Alfred North Whitehead (1861–1947) proposed a type theory in the *Principia Mathematica (1910–1913)* that avoids paradoxes through self-reference.[7] Bertrand Russell explicitly committed

---

[1] See Kleen (1991).

[2] See Dedekind (1880).

[3] Dedekind's construction of the real numbers is fascinating and still relevant today. A real number is identified with a division of the rational numbers into two sets A and B, where all numbers in A are smaller than the numbers in B and A has no greatest element (Dedekind cut). See Dedekind (1880).

[4] Compare Dedekind (1872) and Peano (1889), as well as Sect. 6.4. Since Dedekind's and Peano's introduction of natural numbers is very similar, some authors also speak of the Dedekind-Peano axioms.

[5] See Frege (1884, 1893).

[6] See the correspondence between Frege and Russell in Gabriel et al. (1976).

[7] See Russell and Whitehead (1962) for more details.

134   MATHEMATICAL LOGIC   [PART I

further referred to, unless in cases where its employment is obscure or specially important.

*9·21.   ⊢:. (x).φx ⊃ ψx . ⊃ : (x).φx . ⊃ . (x).ψx

I.e. if φx always implies ψx, then "φx always" implies "ψx always." The use of this proposition is constant throughout the remainder of this work.

Dem.

⊢ . *2·08 .	⊃ ⊢ : φz ⊃ ψz . ⊃ . φz ⊃ ψz	(1)
⊢ . (1) . *9·1 .	⊃ ⊢ : (∃y) : φz ⊃ ψz . ⊃ . φy ⊃ ψz	(2)
⊢ . (2) . *9·1 .	⊃ ⊢ :. (∃x) :. (∃y) : φx ⊃ ψx . ⊃ . φy ⊃ ψz	(3)
⊢ . (3) . *9·13 .	⊃ ⊢ :: (z) :: (∃x) :. (∃y) : φx ⊃ ψx . ⊃ . φy ⊃ ψz	(4)
[(4).(*9·06)]	⊢ :: (z) :: (∃x) :. φx ⊃ ψx . ⊃ : (∃y) . φy ⊃ ψz	(5)
[(5).(*1·01.*9·08)]	⊢ :. (∃x) . ∼(φx ⊃ ψx) : ∨ : (z) : (∃y) . ∼φy ∨ ψz	(6)
[(6).(*9·08)]	⊢ :. (∃x) . ∼(φx ⊃ ψx) : ∨ : (∃y) . ∼φy . ∨ . (z) . ψz	(7)
[(7).(*1·01)]	⊢ :. (x) . φx ⊃ ψx . ⊃ : (y) . φy . ⊃ . (z) . ψz	

This is the proposition to be proved, since "(y).φy" is the same proposition as "(x).φx," and "(z).ψz" is the same proposition as "(x).ψx."

*9·22.   ⊢:. (x).φx ⊃ ψx . ⊃ : (∃x).φx . ⊃ . (∃x).ψx

I.e. if φx always implies ψx, then if φx is sometimes true, so is ψx. This proposition, like *9·21, is constantly used in the sequel.

Dem.

⊢ . *2·08 .	⊃ ⊢ : φy ⊃ ψy . ⊃ . φy ⊃ ψy	(1)
⊢ . (1) . *9·1 .	⊃ ⊢ : (∃z) : φy ⊃ ψy . ⊃ . φy ⊃ ψz	(2)
⊢ . (2) . *9·1 .	⊃ ⊢ :. (∃x) :. (∃z) : φx ⊃ ψx . ⊃ . φy ⊃ ψz	(3)
⊢ . (3) . *9·13 .	⊃ ⊢ :: (y) :: (∃x) :. (∃z) : φx ⊃ ψx . ⊃ . φy ⊃ ψz	(4)
[(4).(*9·06)]	⊢ :: (y) :: (∃x) :. φx ⊃ ψx . ⊃ : (∃z) . φy ⊃ ψz	(5)
[(5).(*1·01.*9·08)]	⊢ :: (∃x) . ∼(φx ⊃ ψx) : ∨ : (y) : (∃z) . φy ⊃ ψz	(6)
[(6).(*1·01.*9·07)]	⊢ :: (∃x) . ∼(φx ⊃ ψx) : ∨ : (y) . ∼φy . ∨ . (∃z) . ψz	(7)
[(7).(*1·01.*9·01·02)]	⊢ :. (x) . φx ⊃ ψx . ⊃ : (∃y) . φy . ⊃ . (∃z) . ψz	

This is the proposition to be proved, because (∃y).φy is the same proposition as (∃x).φx, and (∃z).ψz is the same proposition as (∃x).ψx.

*9·23.   ⊢ : (x).φx . ⊃ . (x).φx   [Id . *9·13·21]

*9·24.   ⊢ : (∃x).φx . ⊃ . (∃x).φx   [Id . *9·13·22]

*9·25.   ⊢ :. (x) . p ∨ φx . ⊃ : p . ∨ . (x) . φx   [*9·23 . (*9·04)]

We are now in a position to prove the analogues of *1·2—·6, replacing one of the letters p, q, r in those propositions by (x).φx or (∃x).φx. The proofs are given below.

---

*9·3.   ⊢ :. (x).φx . ∨ . (x).φx : ⊃ . (x).φx

Dem.

⊢ . *1·2 .	⊃ ⊢ . φx ∨ φx . ⊃ . φx	(1)
⊢ . (1) . *9·1 .	⊃ ⊢ : (∃y) : φx ∨ φy . ⊃ . φx	(2)
⊢ . (2) . *9·13 .	⊃ ⊢ :. (x) :. (∃y) : φx ∨ φy . ⊃ . φx	(3)
[(3).(*9·05·01·04)]	⊢ :. (x) :. φx . ∨ . (y) . φy : ⊃ . φx	(4)
⊢ . (4) . *9·21 .	⊃ ⊢ :. (x) : φx . ∨ . (y) . φy : ⊃ . (x) . φx	(5)
[(5).(*9·03)]	⊢ :. (x) . φx . ∨ . (y) . φy : ⊃ . (x) . φx :. ⊃ ⊢ . Prop	

*9·31.   ⊢ :. (∃x) . φx . ∨ . (∃x) . φx : ⊃ . (∃x) . φx

This is the only proposition which employs *9·11.

Dem.

⊢ . *9·11·13 .	⊃ ⊢ : (y) : φx ∨ φy . ⊃ . (∃z) . φz	(1)
[(1).(*9·03·02)]	⊢ : (∃y) . φx ∨ φy . ⊃ . (∃z) . φz	(2)
⊢ . (2) . *9·13 .	⊃ ⊢ : (x) : (∃y) . φx ∨ φy . ⊃ . (∃z) . φz	(3)
[(3).(*9·03·02)]	⊢ :. (∃x) : (∃y) . φx ∨ φy : ⊃ . (∃z) . φz	(4)
[(4).(*9·05·06)]	⊢ :. (∃x) . φx . ∨ . (∃y) . φy : ⊃ . (∃z) . φz	

*9·32.   ⊢ :. q . ⊃ : (x) . φx . ∨ . q

Dem.

⊢ . *1·3 .	⊃ ⊢ :. q . ⊃ : φx . ∨ . q	(1)
⊢ . (1) . *9·13 .	⊃ ⊢ :. (x) :. q . ⊃ : φx . ∨ . q	
[*9·25]	⊃ ⊢ :. q . ⊃ : (x) : φx . ∨ . q	(2)
[(2).(*9·03)]	⊢ :. q . ⊃ : (x) . φx . ∨ . q	

*9·33.   ⊢ :. q . ⊃ : (∃x) . φx . ∨ . q   [Proof as above]

*9·34.   ⊢ :. (x) . φx . ⊃ : p . ∨ . (x) . φx

Dem.

⊢ . *1·3 .	⊃ ⊢ : φx . ⊃ . p ∨ φx	(1)
⊢ . (1) . *9·13 .	⊃ ⊢ : (x) : φx . ⊃ . p ∨ φx	(2)
⊢ . (2) . *9·21 .	⊃ ⊢ : (x) . φx . ⊃ . (x) . p ∨ φx	(3)
⊢ . (3) . (*9·04) .	⊃ ⊢ . Prop	

*9·35.   ⊢ :. (∃x) . φx . ⊃ : p . ∨ . (∃x) . φx   [Proof as above]

*9·36.   ⊢ :. p . ∨ . (x) . φx : ⊃ : (x) . φx . ∨ . p

Dem.

⊢ . *1·4 .	⊃ ⊢ : p ∨ φx . ⊃ . φx ∨ p	(1)
⊢ . (1) . *9·13·21 .	⊃ ⊢ : (x) . p ∨ φx . ⊃ . (x) . φx ∨ p	(2)
⊢ . (2) . (*9·03·04) .	⊃ ⊢ . Prop	

*9·361.   ⊢ :. (x) . φx . ∨ . p : ⊃ : p . ∨ . (x) . φx   [Similar proof]

*9·37.   ⊢ :. p . ∨ . (∃x) . φx : ⊃ : (∃x) . φx . ∨ . p   [Similar proof]

*9·371.   ⊢ :. (∃x) . φx . ∨ . p : ⊃ : p . ∨ . (∃x) . φx   [Similar proof]

Fig. 7.1 Two pages of the Principia Mathematica

himself to the logicist project in the philosophy of mathematics, at least in his early years.[8] Although to our knowledge type theory does not imply any contradictions, it could not establish itself as the foundation of mathematics. A glance at the Principia Mathematica reveals why this is not surprising, as the developed formalism is very complex, see Fig. 7.1. The more manageable axiom system of set theory by German mathematicians Ernst Zermelo (1871–1953) and Abraham Fraenkel (1891–1965) is now considered the foundation of modern mathematics.[9] Almost all contemporary mathematics, including geometry, can be traced back to this axiom system with some additions and extensions. We will see below that both the system of Russell and Whitehead and the system of Zermelo and Fraenkel contain axioms, which appear evident, but not logically evident. These systems therefore cannot be considered as a complete reduction of mathematics to logic. However, we can say that the reduction of mathematics to logic and set theory has been successful.

---

[8] See the introduction to the Principia Mathematica in Russell and Whitehead (1962) for more details.

[9] See Deiser (2010) for more details.

A revival of logicism in relation to arithmetic came about through the work of British philosopher Crispin Wright (1942–) and American philosopher and logician George Boolos (1940–1996) in the 1980s and 1990s, which is based on a consideration by American philosopher Charles Parsons (1933–).[10] Wright found that Frege's derivation of arithmetic from logic can also be carried out under premises, which are weaker and do not directly lead to contradictions. Formally, this was proven by Boolos and the result is now known as Frege's Theorem, we will discuss this in more detail in Sect. 7.4.[11] This result is the basis of Neologicism, a contemporary logicist-oriented school of thought. Today, there are various neologicist theories, such as modal-logical or intuitionistic neologicism, which differ in their assumptions and the underlying logic. We see Edward Zalta (1952–) as a representative of a modal-logical and Neil Tennant (1950-) as a representative of an intuitionistic direction.[12]

We discuss in Sect. 7.5 whether the assumptions of logicist theories are tenable and can actually be considered logically true, and we will assess whether the logicist project in relation to mathematics as a whole is promising.

7.3 Introduction to Formal Logic

In order to understand the logicist programme of reducing mathematics to formal logic and to highlight the problems of logicism, it is essential to provide an introduction to formal logic. Our goal is to give the reader a basic understanding without providing a comprehensive and detailed introduction to formal logic. For such an introduction, we refer to the relevant literature.[13] We start here with a sketch of propositional logic, to then present elements of predicate logic that are fundamental for a logicist philosophy of mathematics.

A statement in a (two-valued) logic is given by a sentence with a clear truth value w (true) or f (false). If p and q are variables that stand for statements, then:

- $\neg p$ denotes the negation (not p)
- $p \wedge q$ denotes the conjunction (p and q)
- $p \vee q$ denotes the disjunction (p or q)
- $p \Rightarrow q$ denotes the implication (from p follows q)
- $p \Leftrightarrow q$ denotes the equivalence (p if and only if q)

---

[10] See Parsons (1965) for more details.

[11] See Wright (1983) and Boolos (1998) for more details.

[12] See Zalta (1999) and Tennant (2009) as well as Chap. 8 on Intuitionism.

[13] See Rautenberg (2008) or Bohse and Rosenkranz (2006) for humanities scholars.

The operations $\neg, \wedge, \vee, \Rightarrow, \Leftrightarrow$ are called junctors. The truth values of these connections are determined by the following table:

p	q	$\neg p$	$\neg q$	$p \wedge q$	$p \vee q$	$p \Rightarrow q$	$p \Leftrightarrow q$
t	t	f	f	t	t	t	t
t	f	f	t	f	t	f	f
f	t	t	f	f	t	t	f
f	f	t	t	f	f	t	t

Consider, for example, the statements $p =$ "Socrates is a dog" and $q =$ "Socrates is alive", we get:

- $\neg p =$ "Socrates is not a dog"
- $\neg q =$ "Socrates is not alive"
- $p \wedge q =$ "Socrates is a dog and is alive"
- $p \vee q =$ "Socrates is a dog or he is alive"
- $p \Rightarrow q =$ "If Socrates is a dog, then he is alive"
- $p \Leftrightarrow q =$ "Socrates is a dog if and only if he is alive"

If Socrates the philosopher is meant, then the truth value of p and q is false. Thus, $\neg q, \neg p, p \Rightarrow q, p \Leftrightarrow q$ have the truth value true. The other conjunctions have the truth value false. If Socrates is meant to be a living dog, then only $\neg p, \neg q$ are false, the other conjunctions are true.

Expressions obtained by applying the junctors $\neg, \wedge, \vee, \Rightarrow, \Leftrightarrow$ to variables for statements $p, q, r, s, \ldots$ according to certain syntactic rules[14] are called formulas of propositional logic.

A formula α is called logically valid, a tautology or sometimes also a logical truth, if it is true for every assignment of the variables, i.e., it has the value w. The negation of a tautology is a contradiction, i.e., a logical contradiction. Tautologies of (two-valued) propositional logic include the law of non-contradiction $\neg(a \wedge \neg a)$, the law of excluded middle $\neg a \vee a$, the modus ponendo ponens

$$(a \wedge (a \Rightarrow b)) \Rightarrow b$$

or the contraposition

$$(a \Rightarrow b) \Leftrightarrow (\neg b \Rightarrow \neg a).$$

Tautologies play a special role for logicism in the philosophy of mathematics. The logicist claim in its strongest form is that all statements of mathematics can be

---

[14] We refrain from introducing the grammar of a logical language here and refer to Rautenberg (2008).

reduced to tautologies with suitable definitions. It is obvious that propositional logic is not sufficient for the logicist project, so we now provide an introduction to predicate logic.

A unary predicate is a statement form $A(x)$ that contains a variable x, such that when the variable is replaced by elements from a given individual domain U, also called a universe, a statement with a uniquely determined truth value is obtained. An n-ary predicate is a statement form $A(x_1, \ldots, x_n)$ with the variables $x_1, \ldots x_n$, such that when all variables are replaced by elements of an individual domain, a statement is obtained. Statement forms with two variables are also called (binary) relations.

The existential quantifier $\exists$ transforms a unary predicate $A(x)$ into an existence statement. That is, $\exists x : A(x)$ is true if and only if $A(x)$ is true for at least one individual x from U, otherwise the statement is false. The universal quantifier $\forall$ transforms a unary predicate into a universal statement. That is, $\forall x : A(x)$ is true if and only if $A(x)$ is true for all individuals from U and otherwise false. Sometimes the quantifier $\exists! x : A(x)$ is used, which states that exactly one x exists such that $A(x)$ is true. Formally, this quantifier can be defined by the existential quantifier and universal quantifier in the following way:

$$\exists! x : A(x) \Leftrightarrow \exists x : (A(x) \wedge \forall y : A(y) \Rightarrow y = x).$$

A quantifier binds a variable in an n-ary predicate $A(x_1, \ldots, x_n)$ and generates an $n - 1$-ary predicate. If each variable is bound by n quantifiers, a statement is obtained.

We consider as an example the binary predicate $F(x, y) = x$ loves y. The universe consists of all people and includes in particular Alice and Peter. The following expressions are unary predicates, but not statements:

- $F(\text{Peter}, y) =$ "Peter loves y"
- $F(x, \text{Alice}) =$ "x loves Alice"
- $\forall x : F(x, y) =$ "Everyone loves y"
- $\exists x : F(x, y) =$ "There is at least one person who loves y"
- $\exists! x : F(x, y) =$ "There is exactly one person who loves y"
- $\forall y : F(x, y) =$ "x loves everyone"
- $\exists y : F(x, y) =$ "x loves at least one person"
- $\exists! y : F(x, y) =$ "x loves exactly one person"

The following expressions are statements:

- $F(\text{Peter}, \text{Alice}) =$ "Peter loves Alice"
- $\forall x : F(x, \text{Alice}) =$ "Everyone loves Alice"
- $\exists x : F(x, \text{Alice}) =$ "At least one person loves Alice"
- $\exists! x : F(x, \text{Alice}) =$ "Exactly one person loves Alice"
- $\forall y : F(\text{Peter}, y) =$ "Peter loves everyone"
- $\exists y : F(\text{Peter}, y) =$ "Peter loves at least one person"

- $\exists! y : F(\text{Peter}, y) = $ "Peter loves exactly one person"
- $\forall x \forall y : F(x, y) = $ "Everyone loves everyone"
- $\exists x \exists y : F(x, y) = $ "There is at least one person who loves at least one person"
- $\forall x \exists y : F(x, y) = $ "Everyone loves at least one person"
- $\forall y \exists x : F(x, y) = $ "Everyone is loved by at least one person"

Expressions obtained by applying junctors and quantifiers according to certain syntactic rules to statement forms are called We form formulas of predicate logic.[15] Such a formula is a tautology if it is true for every model, i.e. all universes and propositional forms. Examples of tautologies in predicate logic are $\forall x : A(x) \Rightarrow A(a)$ or $A(a) \Rightarrow \exists x : A(x)$, if a is in the domain of individuals. Further tautologies connect quantifiers and junctors:

$$\neg(\forall x : A(x)) \Leftrightarrow \exists x : \neg A(x), \quad \neg(\exists x : A(x)) \Leftrightarrow \forall x : \neg A(x),$$

$$(\exists x : A(x)) \vee (\exists x : B(x)) \Leftrightarrow \exists x : A(x) \vee B(x)$$

$$(\forall x : A(x)) \wedge (\forall x : B(x)) \Leftrightarrow \forall x : A(x) \wedge B(x).$$

In first-order predicate logic, we do not consider propositional forms $A(x)$ that refer to predicates from the domain of individuals U, or to sets of individuals, but only propositional forms that have individuals as arguments. In first-order predicate logic, we therefore only quantify over individuals. In second-order predicate logic, we also consider propositional forms $A(x)$, which can have predicates on a domain of individuals or sets of individuals as arguments in addition to individuals x. Quantifying over such objects is allowed in second-order predicate logic. Second-order predicate logic is more expressive than first-order predicate logic; in it, statements can be formulated that cannot be formulated in first-order predicate logic. An important example is the original form of the induction principle of arithmetic, which we introduce in Sect. 7.4. Examples of tautologies in second-order predicate logic, in which quantification is over predicates, are:

$$\forall x \forall A : \neg(A(x) \wedge \neg A(x)), \qquad \forall x \forall A : A(x) \vee \neg A(x)$$

$$\forall x (\neg \exists A : A(x) \Leftrightarrow \forall A : A(x)).$$

Second-order predicate logic is a genuine generalisation of first-order predicate logic, which has the disadvantage that some sentences in first-order predicate logic, such as the compactness theorem or Gödel's completeness theorem, lose their validity. Logicism in the philosophy of mathematics generally assumes second-order predicate logic in the attempt to reduce mathematics, and in particular arithmetic, to logic.

---

[15] We refrain from introducing the grammar of predicate logic, and refer again to Rautenberg (2008). It should only be noted that quantifiers generally bind stronger than junctors, so they do not need to be bracketed.

7.4   The Attempt to Reduce Arithmetic

The Peano axioms are usually accepted by mathematicians and philosophers as the basis of arithmetic. An exception to this are the proponents of finitism in the philosophy of mathematics, who do not want to allow the infinite in any form.[16]
The Peano axioms for the natural numbers $\mathbb{N}$, including zero, are as follows:[17]

1. 0 is a natural number.
   ($0 \in \mathbb{N}$)
2. Every natural number n has a natural number $N(n)$ as a successor.
   ($\forall n \exists N(n) : n \in \mathbb{N} \Rightarrow N(n) \in \mathbb{N}$)
3. There is no natural number whose successor is 0.
   ($\forall n : n \in \mathbb{N} \Rightarrow N(n) \neq 0$)
4. Two different natural numbers n, m have different successors.
   ($\forall n, m : (n, m \in \mathbb{N} \wedge n \neq m) \Rightarrow N(n) \neq N(m)$).
5. If A is a unary predicate on the natural numbers, then: If $A(0)$ is true and the truth of $A(n)$ implies the truth of $A(N(n))$ for all natural numbers, then $A(n)$ is true for all natural numbers.
   ($\forall A(A(0) \wedge (\forall n \in \mathbb{N} : A(n) \Rightarrow A(N(n))) \Rightarrow \forall n \in \mathbb{N} : A(n))$)

The last axiom describes the principle of complete induction. This principle is not only the basis for induction proofs and recursive definitions in arithmetic, but in all of mathematics. The classical formulation given here obviously requires second-order predicate logic. As an axiom schema it can, however, also be formulated in first-order predicate logic. Instead of predicates A, we consider formulas $\phi(n)$ in first-order predicate logic, which contain natural numbers as free arguments. For each of these formulas, we now add the axiom

$$\phi(0) \wedge (\forall n \in \mathbb{N} : \phi(n) \Rightarrow \phi(N(n))) \Rightarrow \forall n \in \mathbb{N} : \phi(n))$$

to our axiom system. The disadvantage of this system is that it contains an infinite number of axioms that cannot be explicitly formulated in their entirety.
The concern of logicism with respect to arithmetic is to derive the Peano axioms from logically evident premises and suitable definitions. We initially assume second-order predicate logic and consider, in addition to a predicate A, its scope U. This is the class of objects to which the predicate applies, i.e. $U = \{x \,|\, A(x)\}$, and is also referred to as the extension of the predicate. Gottlob Frege bases his classical derivation of the Peano axioms on the following axiom in second-order predicate logic about the scope of predicates:

$$\forall A, B : \{x \,|\, A(x)\} = \{x \,|\, B(x)\} \Leftrightarrow \forall x : A(x) \Leftrightarrow B(x),$$

---

[16] See also Chap. 8 on intuitionism and Chap. 10 on constructivism.
[17] See Peano (1889).

for all predicates A and B. This is Basic Law V from Frege's *Basic Laws of Arithmetic* in modern notation.[18] It means that the scope of two predicates, i.e., the classes of objects to which they apply, is exactly the same when the predicates are equivalent, i.e., the first predicate applies to an object exactly when the second predicate applies to this object. Frege's derivation of arithmetic is complex and is considered conclusive by experts. Performing the derivation is beyond the scope of this book.[19] Frege's work would be a great success for logicism in the philosophy of mathematics, if Basic Law V were indeed logically evident, and in particular true. This is not the case.

Given an arbitrary predicate A and applying the Basic Law with $A = B$, we get $\{x|A(x)\} = \{x|A(x)\}$, since $\forall x : A(x) \Leftrightarrow A(x)$ is a tautology. In particular, the scope exists for every predicate, i.e.

$$\forall A \exists U : U = \{x|A(x)\},$$

or equivalently formulated

$$\forall A \exists U \forall x : x \in U \Leftrightarrow A(x).$$

This necessary consequence of Basic Law V is generally false, as it leads to a contradiction. To see this, consider the predicate $A(x)$, which states that the class x does not contain itself as an element, i.e. $A(x) \Leftrightarrow x \notin x$. Now applying the existence of the scope U to this predicate, we get

$$\forall x : x \in U \Leftrightarrow x \notin x.$$

Setting $x = U$, we get $U \in U \Leftrightarrow U \notin U$, i.e., U contains itself as an element exactly when U does not contain itself as an element, a contradiction. This is the famous Russell's paradox, which shows that not every predicate consistently defines a scope, in the sense of the set of objects to which the predicate applies. This finding marks the end of the naive use of the concept of set as a summary of objects that have a property. We already mentioned in Sect. 7.3 that Russell and Whitehead developed a type theory and Zermelo and Fraenkel developed an axiomatic set theory to avoid contradictions like in Russell's paradox. Both systems are complex and extensive. We refrain from presenting the type theory here and refer to the appendix (Chap. 14) of this book for an introduction to the set theory of Zermelo and Fraenkel. Here we only present some properties of these theories that are relevant to logicism in relation to arithmetic. Indeed, the Peano axioms can be derived from both theories, so a reduction of arithmetic to a more fundamental theory is possible. In contrast to Frege's Basic Laws, no contradictions have been derived from the type theory and the axiomatic set theory to our knowledge. On the other hand, neither

---

[18] See Frege (1893).

[19] See Heck (2012) for a contemporary presentation.

of these formal systems can prove their own consistency. This is a consequence of the second incompleteness theorem of the Austrian-American mathematician Kurt Gödel (1906–1978). This states that no consistent formal system that contains elementary arithmetic can prove its own consistency.[20] In relation to the logicist project, it is also important to note that both the type theory and the axiomatic set theory postulate the existence of infinitely many individuals of the first type, or an infinite set, respectively. In the language of set theory, this axiom has the following form:

$$\exists A \colon (\emptyset \in A \wedge \forall x \colon (x \in A \Rightarrow x \cup \{x\} \in A)),$$

i.e., there is a set A that contains the empty set $\emptyset$ and with every element x also the union of x and $\{x\}$. Writing A in the form of a listing, we get

$$A = \{\emptyset, \ \{\emptyset, \{\emptyset\}\}, \ \{\emptyset, \ \{\emptyset, \{\emptyset\}\}\}, \ \{\emptyset, \ \{\emptyset, \{\emptyset\}\}, \ \{\emptyset, \ \{\emptyset, \{\emptyset\}\}\}\}, \ldots \}$$

It is basically not surprising that the Peano axioms can be derived from this assumption by making the following definition:

$$0 := \emptyset \qquad N(n) := n \cup \{n\},$$

where $N(n)$ is the successor of a natural number n. Whether the reduction of arithmetic outlined here, or any other approach that presupposes an axiom of infinity, can be understood as a reduction in the sense of logicism, we will discuss in Sect. 7.5.

We will present another attempt at reducing arithmetic here, which is strongly based on Frege's work and forms the basis of the neo-logicist philosophy. This approach again starts from a second-order predicate logic and continues to base itself on Hume's Principle. Informally, Hume's Principle states:

The number of Fs is equal to the number of Gs exactly when there is a one-to-one correspondence between the Fs and the Gs.

This principle goes back to the work *A Treatise of Human Nature* by the Scottish philosopher David Hume (1711–1776).[21] To formalise the principle, we first need to define for two predicates F and G in purely logical terms what it means for there to be a one-to-one correspondence between the objects that satisfy F and the objects that satisfy G. This is formally done by the following expression

$$\exists R : (\forall x (F(x) \Rightarrow \exists! y : (G(y) \wedge R(x, y))) \wedge \forall y (G(y) \Rightarrow \exists! x : (F(x) \wedge R(x, y)))),$$

which states: There exists a relation (a binary predicate) R, such that every object to which F applies is in relation to exactly one object to which G applies, and every

---

[20] See Nagel and Newman (2003) for a proof.

[21] See Hume (2000).

object to which G applies is in relation to exactly one object to which F applies. If such a relation exists, we say that F and G are equinumerous, and write this as $F \approx G$. It is easy to see that this is an equivalence relation for predicates, i.e., it holds

$$F \approx F, \qquad F \approx G \Leftrightarrow G \approx F, \qquad F \approx G \wedge G \approx H \Rightarrow F \approx H,$$

for all F, G and H. Hume's Principle now states

$$\sharp F = \sharp G \Leftrightarrow F \approx G,$$

where $\sharp F$ denotes the number of objects that satisfy F, and $\sharp G$ denotes the number of objects that satisfy G. Implicitly, Hume's Principle defines *the number of objects that satisfy F* as the equivalence class of F with respect to the equivalence relation $\approx$, i.e., $\sharp F$ includes all predicates that are equivalent to F with respect to $\approx$. The definition of equinumerosity of predicates given here is similar to the definition of equinumerosity of sets by the German mathematician Georg Cantor (1845–1918). Two sets have the same power or cardinality if there is a one-to-one mapping between them. Crucial for logicism, however, is that we do not rely on the concept of set here.

From Hume's Principle and a second-order predicate logic, the Peano axioms can be derived. The proof of this theorem essentially goes back to Frege's work and the theorem is therefore called Frege's Theorem. Frege provides a derivation of Hume's Principle from the untenable Basic Law V discussed above and then carries out the derivation of the Peano axioms using Hume's Principle. To formulate this derivation exceeds the scope of this book.[22] We only want to indicate how 0 and the successor $N(n)$ of a natural number n can be determined starting from Hume's Principle. On the one hand, we define $\sharp F = 0$ exactly when F does not apply to any object, i.e.

$$\sharp F = 0 \; :\Leftrightarrow \neg \exists x : F(x).$$

On the other hand, $N(n)$ is the immediate successor of a natural number n exactly when there is a predicate F and an object w to which F applies, such that $N(n)$ is the number of objects to which F applies, and n is the number of objects other than w to which F applies. Formally, this means:

$$\exists F \exists w : F(w) \wedge N(n) = \sharp F \wedge n = \sharp(F(x) \wedge x \neq w).$$

It should also be mentioned here that Hume's Principle can be added to a second-order predicate logic without contradiction, such a logic is consistent exactly when it is consistent extended by Hume's Principle. So far, we have outlined the basic approach of Neo-Logicism. In current neo-logicist works, however, Hume's

---

[22] We refer again to Frege (1893) and Heck (2012).

Principle is no longer assumed in its general form, but is restricted. Predicates are distinguished according to whether they apply to concrete or abstract objects, and Hume's Principle is only asserted for predicates, which apply to concrete objects. Thus, the number of objects $\sharp F$ to which a predicate F applies is only defined for predicates, which apply to concrete objects, and not for arbitrary predicates.[23]

Whether the reduction of arithmetic to a second-order predicate logic extended by Hume's Principle outlined here can be understood as a reduction of mathematics to logic in the sense of the logicist project, we discuss in the following section.

7.5 Assessment of the Logicist Endeavour

In this section, we undertake a critical assessment of the logicist endeavour to reduce mathematics to logic. First, we address the question to what extent the attempts to reduce arithmetic, which were presented in Sect. 6.4, are viable. Afterwards, we argue against the possibility of reducing all of mathematics to logic.

Let us first consider the reduction of arithmetic to the type theory of Russell and Whitehead or the set theory of Zermelo and Fraenkel. Both approaches seem successful, but they both require, among other things, the existence of infinitely many objects of the first type, or an infinite set through an infinity axiom. A justified assessment of such an axiom is easy, it is not logically evident within a type theory or set theory. The talk of objects of a type or sets of objects does not commit us to the claim that there are infinitely many objects of a type or infinite sets. This argument can be made precise. The Zermelo-Fraenkel set theory is consistent with the infinity axiom exactly when it is consistent with the assumption of the negation of the infinity axiom.[24] The assumption that there is no infinite set can only be refuted within the axiomatic set theory, if the set theory, regardless of which assumption about the existence of an infinite set is made, is contradictory. The same applies to the assumption that there is an infinite set. A similar argument can also be formulated for the type theory according to Russell and Whitehead. We therefore see that the widely recognised foundation of mathematics and in particular arithmetic does include a predicate logic, but does not represent a reduction of mathematics to a predicate logic. Logicism was indeed a major driving force for the formalisation of the foundations of arithmetic through type theory or set theory, but these formalisations contain axioms that are not tautologies in the sense of predicate logic.

Let us now consider the justification of arithmetic through a second-order predicate logic together with Hume's Principle, as proposed by neologicism. This stands or falls with the validity of Hume's Principle. Interpreting and assessing Hume's Principle is significantly more complex than assessing the axiom of infinity.

---

[23] We find such an approach in Zalta (1999), where a modal logic is assumed.
[24] See Kunen (1980).

For a predicate F, $\sharp F$ denotes the number of objects that satisfy F. If we understand *number of objects* to mean their cardinality, i.e., a cardinal number, Hume's Principle takes the following form:

The cardinal number of objects that satisfy F is equal to the cardinal number of objects that satisfy G if there is a one-to-one correspondence between the objects that satisfy F and the objects that satisfy G.

This is nothing other than the definition of the identity of cardinal numbers, as used in set theory. Therefore, Hume's Principle can be seen as logically evident in relation to cardinal numbers, insofar as the cardinality of the objects that fall under the predicates F and G exists. Hume's Principle in its present form implies that for a predicate F there exists a cardinal number $\sharp F$, which denotes the cardinality of the objects that fall under F. This conclusion from Hume's Principle seems problematic to us for two reasons for a logicism.

For a predicate F, $\sharp F$ is defined by the equivalence relation of equinumerosity $\approx$ and therefore denotes the equivalence classes of predicates equinumerous to F. The ontological question arises in what sense this equivalence class exists. The common mathematical interpretation is that the equivalence class of F is the set of objects that are equivalent to F, i.e., $\sharp F = \{G | G \simeq F\}$. At this point, we suspect that neologicism implicitly presupposes a set theory and we question whether the set theory needed in this context can actually do without axioms that are not logically evident. At least we find no satisfactory answer in the neologicist literature to the question of what the term $\sharp F$, which is not part of a predicate logic, should refer to.

The second problem is that Hume's Principle in its general form implies the existence of a cardinal number for every predicate F. Consider the predicate $x = x$ of self-identity, which all objects satisfy. From Hume's Principle it follows that this predicate has a cardinal number, that is, the cardinality of all objects in a second-order predicate logic exists. This number is sometimes called anti-zero or universal number in the literature. Logicians are aware that the assumption of the existence of such a number is highly questionable. In response to problems with Hume's Principle in its general form, attempts are made to restrict the principle to a suitable set of predicates. In the contemporary literature, for example, the distinction between predicates that refer to ordinary objects and predicates that refer to abstract objects is proposed.[25] We cannot conclusively assess these approaches here, but we would like to express doubts that the necessary distinction of predicates can be ensured by purely logical criteria.

Neologicism interprets, based on Hume's Principle, natural numbers as cardinal numbers. At this point, we would like to address the question of whether this interpretation is adequate. In the set-theoretic formalism, natural numbers can also be understood as ordinal numbers, which are uniquely determined by their position in a suitable order:

$$0 := \emptyset < 1 := \{0\} < 2 := \{0, 1\} < 3 := \{0, 1, 2\} < 4 := \{0, 1, 2, 3, \} < \ldots$$

---

[25] See, for example, Zalta (1999).

The concept of finite or finite ordinal numbers can be extended in a straightforward way to transfinite ordinal numbers:

$$\forall n \in \mathbb{N} : n < \omega := \{0, 1, 2, 3, \dots\} < \omega + 1 := \{0, 1, 2, 3, \dots, \omega\}$$

$$< \omega + 2 =: \{0, 1, 2, 3, \dots, \omega, \omega + 1\} < \cdots < \omega + n$$

$$:= \{0, 1, 2, 3, \dots, \omega, \dots, \omega + (n - 1)\}$$

$$\cdots < 2\omega := \{0, 1, 2, 3, \dots \omega + 1, \omega + 2, \dots\}.$$

For transfinite ordinal numbers, Hume's Principle obviously does not hold. There are one-to-one correspondences between ω, $\omega + 1$, $\omega + 2$, 2ω etc., although these numbers are clearly distinguished by their position in the order. We leave it to the reader to provide such correspondences. Transfinite arithmetic cannot be derived from a second-order predicate logic, together with Hume's Principle. This is not significant for the logical derivation of the Peano axioms from Hume's Principle using a predicate logic. However, one might hold the view that the interpretation of natural numbers as cardinal numbers, which neologicism propagates, is inadequate. This interpretation does not allow for generalisations of natural numbers to transfinite numbers. For Georg Cantor (1845–1918), who developed the theory of ordinal and cardinal numbers along with set theory, ordinal numbers were probably the basis of arithmetic. In any case, transfinite numbers are today an indispensable part of modern arithmetic. An example that illustrates this is the theory of Goodstein sequences.[26]

So far, we have critically engaged with the attempt to reduce arithmetic to logic. We now present two more arguments that highlight further problems of logicism in the philosophy of mathematics. The first argument is based on the assumption of the validity of the axiom of choice and includes the exclusion of the reduction of some central theorems of modern mathematics to logic. The second argument is based on Gödel's first incompleteness theorem and shows that the reduction of all true mathematical theorems to logic is impossible.

The axiom of choice states that for a non-empty set of sets there is a selection function that assigns one of its elements to each of these sets. Formally expressed, this means:

$$\forall X \left(\emptyset \notin X \Rightarrow \exists f : X \to \bigcup_{A \in X} A \quad : \quad \forall A \in X : (f(A) \in A) \right)$$

The axiom of choice is an addition to the axiomatic set theory of Zermelo and Fraenkel. Based on the validity of this axiom, the following argument can be

---

[26] See Chapter 1 in Neunhäuserer (2015).

formulated:

(1) The axiom of choice does not follow from logically evident premises.
(2) Some central theorems of mathematics are equivalent to the axiom of choice.

From (1) and (2) it follows: Some central theorems of mathematics do not follow from logically evident premises, so their reduction to logic is excluded.

Regarding the first premise of the argument, it should first be noted that we are not aware of any approach that attempts to derive the axiom of choice from logically evident premises. However, a result of the American logician and mathematician Paul Cohen (1934–2007) is more significant: The axiomatic set theory, extended by the axiom of choice, is consistent if and only if it is consistent when extended by the negation of the axiom of choice.[27] It is therefore certain that the axiom cannot be derived from the axioms of Zermelo-Fraenkel set theory together with a predicate logic. To justify the second premise of the argument, the following central theorems of modern mathematics, which are equivalent to the axiom of choice, are mentioned:

- The product of a non-empty family of non-empty sets is not empty. (Set theory)
- Every set can be well-ordered. (Order theory)
- Every non-empty partially ordered set, in which every chain has an upper bound, contains a maximal element. (Order theory)
- Every vector space has a basis. (Algebra)
- The product of compact spaces is compact. (Topology)
- Every undirected, connected graph has a spanning tree. (Graph theory)

In addition, many central theorems in almost all areas of modern mathematics, starting from Zermelo-Fraenkel set theory, cannot be proven without the axiom of choice. Examples of such theorems are:

- Every union of countable sets is countable. (Set theory)
- Every number field has an algebraic closure. (Algebra)
- The additive groups of real and complex numbers are isomorphic. (Algebra)
- A space is compact if and only if it is complete and totally bounded. (Topology)
- Existence of the Lebesgue measure (Analysis)
- The Hahn-Banach theorem (Functional analysis)

Although these theorems are weaker than the axiom of choice, no proof from logically evident premises is known for these theorems either. It is not expected that these theorems will be provable from logically evident premises of any kind. This is clearly a serious problem for logicism with respect to large parts of contemporary mathematics.

---

[27] See Cohen (1966) for the proof of this result.

Finally, we present an argument against the possibility of a complete reduction of mathematics to logic, based on the first incompleteness theorem of Kurt Gödel (1906–1978). This states that any consistent formal system that includes the arithmetic of natural numbers is incomplete, i.e., there are true statements that can be expressed in the language of the formal system but are not provable[28] We formulate the following argument based on this result:

(1) A mathematical theory can be completely reduced to logic if and only if all theorems (true statements) of the theory can be derived from logically evident premises by means of logical deduction.
(2) Concepts of a mathematical theory, logically evident premises and logical inference rules can be described by a formal system in the sense of Kurt Gödel.

From (1) and (2) it follows: (3) A mathematical theory can be completely reduced to logic if and only if all true statements of the theory can be expressed and proven in a formal system.

From (3) and Gödel's incompleteness theorem it follows: A mathematical theory that includes the arithmetic of natural numbers, cannot be completely reduced to logic.

The first premise of the argument is nothing more than a further explanation of our concept of the reduction of a mathematical theory to logic. The second premise of the argument asserts the formalisability of mathematical theories and logical systems. Indeed, the successes of such a formalisation are great, see also Chap. 9 on formalism in the philosophy of mathematics. It seems uncontroversial that large parts of mathematics and logic, in particular arithmetic and predicate logic, can be expressed in a suitable formal language. This speaks for the second premise of the argument. The conclusion of the argument is commonly accepted in the philosophy of mathematics today. We therefore see that logicism in the philosophy of mathematics, insofar as it asserts the possibility of a complete reduction of mathematics to logic, can be considered as failed.

---

[28] See Nagel and Newman (2003).

Intuitionism 8

Contents

8.1   Luitzen Brouwer

Intuitionism in the philosophy of mathematics is attributed to the Dutch mathematician Luitzen Brouwer (1881–1966) and is to this day inextricably linked with his work. We therefore begin this chapter with a brief biography.[1]

Luitzen Brouwer was born in 1881 near Rotterdam. After attending school in Hoorn and Haarlem, he graduated from high school in 1897 and studied at the Faculty of Mathematics and Natural Sciences at the University of Amsterdam. In 1906, Brouwer published his first book *Life, Art and Mysticism* on philosophical and moral issues.[2] This book already reflected Brouwer's subjectively idealistic attitude, which also characterises his philosophy of mathematics. Brouwer received his doctorate in 1907 from the University of Amsterdam on the foundations of mathematics. His dissertation deals with the difference between mathematics and logic and contains the first approaches to intuitionism in the philosophy of mathematics. In a subsequent article in 1908, Brouwer first doubted the law of excluded middle of classical logic, which states that a statement is either true or

---

[1] See van Dalen (1999) for more information.

[2] See Brouwer (1905).

J. Neunhäuserer, *Introduction to the Philosophy of Mathematics*,
Mathematics Study Resources 22, https://doi.org/10.1007/978-3-662-72179-7_8

false.[3] Brouwer then focused intensively on mathematical issues in the field of topology. His most famous result is probably Brouwer's fixed point theorem, which states that a continuous mapping of the closed unit sphere onto itself has a fixed point. In addition, Brouwer generalised Jordan's curve theorem to n dimensions and clarified the concept of topological dimension.[4] In 1912, he was called to a professorship at the University of Amsterdam, where Brouwer taught until his retirement in 1952. In the 1920s and 1930s, Brouwer refined the intuitionistic philosophy of mathematics and developed a set theory, a theory of real numbers, and a function theory, based on the principles of intuitionism. He gave lectures on intuitionism at renowned universities and opposed the progressive formalisation of mathematics, as pursued by David Hilbert (1862–1943) and his students in Göttingen.[5] Hilbert and Brouwer are the two main opponents in the so-called foundational dispute between formalism and intuitionism, which led to a large number of publications in the field of foundations and philosophy of mathematics. As a consequence of this dispute, Hilbert excluded Brouwer from the co-editorship of the Mathematical Annals in 1928, which was the most important mathematical journal at the time. Brouwer had previously published some of his most important works in the Mathematical Annals. After the Second World War, Brouwer seems to be quite isolated in the international mathematical community. He was more concerned with metaphysical than mathematical issues. From this period comes, among other things, Brouwer's concept of the creative mathematical subject, which we discuss in Sect. 8.1. Although Brouwer received numerous awards for his work, such as membership in the Royal Dutch Academy of Sciences and the Royal Society of London, as well as an honorary doctorate from the University of Cambridge, his life seems to have been marked by grievances and conflicts. Brouwer died at the age of 85 in a car accident and left no children.

The further development of intuitionism was taken over by Arend Heyting (1898–1980), a student of Brouwer's, and Dirk van Dalen (1932–), who received his doctorate from Heyting. In particular, a formal interpretation of intuitionistic logic goes back to Heyting.[6] Independently of Heyting, the famous Russian mathematician Andrey Kolmogorov (1903–1987) also presented such an interpretation. In Sect. 8.3, we provide an introduction to intuitionistic logic, based on these interpretations, and then present some elements of intuitionistic mathematics. The chapter concludes with a critique of intuitionism (Sect. 8.5).

---

[3] See Brouwer (1908).

[4] See Brouwer (1911, 1910, 1911).

[5] See also Chap. 9 on formalism.

[6] See Heyting (1930) for more information.

8.2   The Metaphysics of Intuitionism

Fundamental to intuitionism in the philosophy of mathematics is the assumption that mathematics is a creation of the mind. The objects of mathematics, such as numbers, therefore do not exist independently of mental processes and they are not abstract. A mathematical object is not timeless, as it only begins to exist through a creative mental act. According to intuitionism, a mathematical statement being true means that there is a mental process that shows that the statement is true. For the intuitionist, the communication of mathematical content serves to trigger one's own mental process in the mind of another person.

We will explain the basic idea of intuitionism in this section and show in Sect. 8.3 that this has significant consequences for logic and mathematics. First, however, a note on nomenclature is appropriate. The claim that mathematics is a creation of the mind could understandably be referred to as creationism in the philosophy of mathematics, as creation and creation are synonymous terms. Now, the term *creationism* is religiously charged and this connotation is undesirable in our context. The use of the term *intuitionism* is also problematic and can lead to misunderstandings. In the philosophy of mind, an intuition is not a creative act, but an immediate insight. It is usually assumed that such an insight refers to objects that lie beyond the mental process of insight. We would like to emphasise that intuitionism in the philosophy of mathematics cannot be understood in this sense. In contrast to intuitive Platonism, it explicitly rejects the idea that mathematical intuition refers to non-mental objects, see Chap. 3 for comparison on Platonism in the Philosophy of Mathematics.

The concept of the mind is ambiguous and complex in philosophy. To understand intuitionism in the philosophy of mathematics, we must first clarify which conception of the mind and which conception of mental processes is being assumed. Brouwer considers the creative mathematical subject as an ideal mind, which is free from constraints by space and time and the possibility of errors and mistakes.[7] The mental process of mathematical creation is not a cognitive process of a person; it is impersonal and universal. Mathematics is the product of the one ideal creative subject, in which a mathematician participates. Brouwer's creative subject can be understood as a transcendental subject in the sense of the Austrian philosopher Edmund Husserl (1859–1938).[8] By transcendental subject, we mean the I of pure consciousness, which is permanent beyond the contents of consciousness. The metaphysics underlying intuitionism can therefore be described as subjectivist and transcendental idealism. We believe we recognise here motives of German idealism and in particular the philosophy of Gottlieb Fichte (1762–1814).

Brouwer's intuitionism assumes that the mind constitutes mathematics through two pre-linguistic activities, which only presuppose time as an a priori principle. The perception of the passing of time, through which each moment falls into two

---

[7] See Brouwer (1948).

[8] See for this interpretation van Atten (2007).

different parts, what was, and what is together with what was, creates twoness. From there, threeness etc. results from what was, and what is together with what was. Brouwer calls this first act of the subject the primal intuition. It forms the basis of arithmetic in intuitionism and is reminiscent of Kant's philosophy of mathematics.[9] The second act of the creative subject, which underlies mathematics, consists in the free selection of the entries of a sequence of mathematical objects from previously created mathematical objects. So here it is free decisions of the mind that create mathematics. The second act of the creative subject is the basis for the construction of real numbers and thus the basis of analysis in intuitionism. We will go into this in more detail in Sect. 8.4.

The two basic principles of intuitionism mentioned above illustrate the difference between intuitionism and Platonic as well as formalistic approaches in the philosophy of mathematics. Intuitionism neither assumes the existence of a non-mental mathematical reality, as Platonism does, nor does it consider mathematics as a game with symbols according to fixed rules, as formalism does.[10] Intuitionism also differs from logicism, as it assumes a time-dependent concept of truth for mathematical statements. A proven mathematical statement is true at the time and in the future of the time at which it was proven; the statement had no truth value before it was proven. This time dependence can be specified by three axioms.

For all mathematical statements a the following holds:

(1) The creative subject has a proof at a time n that a is true, or the creative subject has no proof at a time n that a is true.
(2) If the creative subject has a proof at a time n that a is true, it also has a proof at all times $m \geq n$ that a is true.
(3) a is true if and only if there is a time n at which the creative subject has a proof that a is true.

The fact that the creative subject has a proof for a mathematical statement a at a time n, means nothing else in intuitionism than that the subject is undergoing a mental process at time n that demonstrates the truth of a, or remembers such a process. The first axiom is not a law of excluded middle for statements a, it merely states that the subject knows at a time that a is true, or does not know this. The second axiom describes the creative subject as an ideal mind that never forgets a proof. The third and perhaps most important axiom defines the truth of a mathematical statement as its provability. Classical logic does not provide for such an identification of truth and provability. In classical logic, every statement a has a unique truth value true or false, regardless of whether a proof for a or a proof for the negation of a is ever found. In Sect. 8.3 we will present the differences between classical and intuitionistic logic, which result from the metaphysical basic assumptions of intuitionism outlined above.

---

[9] See Chap. 5 for this.

[10] See Chaps. 2 and 7 for this.

8.3 Intuitionistic Logic

As stated in Sect. 8.2, a mathematical statement in intuitionism is true exactly when we as a creative subject have or will have a proof of the statement. As long as we have not found a proof of a statement or its negation, we do not know whether the statement is true or false. The identification of truth and provability is fundamental to the logic of intuitionism. The meaning of the junctors and quantifiers is usually determined in intuitionistic logic by the Brouwer-Heyting-Kolmogorov interpretation (BHK interpretation).[11] This interpretation is sometimes also called proof interpretation and consists from the following stipulations:

For all mathematical statements a, b and all mathematical statement forms A the following holds:

(1) A proof of $a \vee b$ is given by a proof of a or a proof of b.
(2) A proof of $a \wedge b$ is given by a proof of a and a proof of b.
(3) A proof of $a \Rightarrow b$ is a construction that allows us to transform any proof of a into a proof of b.
(4) A contradiction $\bot$, is a statement that has no proof. (Often $1 = 0$ is used).
(5) A proof of $\neg a$ is a proof that a implies a contradiction, i.e. $a \Rightarrow \bot$.
(6) A proof of $\forall x : A(x)$ is a construction that allows us to transform any proof of $d \in D$ into a proof of $A(d)$. (D is the domain over which we quantify.)
(7) A proof of $\exists x : A(x)$ consists in specifying a $d \in D$ and a proof of $A(d)$. (D is again the domain over which we quantify.)

The BHK interpretation of intuitionistic logic is not a formal definition, as it presupposes the concept of construction, but does not define it. This concept is indeed understood in different ways and this has an influence on which formulas are logically evident. We will discuss this in Chap. 10 on constructivism in the philosophy of mathematics. The intuitionistic logic in the BHK interpretation can be understood as the basis of the logic of constructivism in the philosophy of mathematics. Some authors therefore consider intuitionism as a variant of constructivism.

The most important difference between classical and intuitionistic logic is that the law of excluded middle a or not a, i.e. $a \vee \neg a$, does not hold in intuitionistic logic. This is not significantly affected by how the concept of construction in the above definition is understood. We would like to clarify this fundamental difference. $a \vee \neg a$ means, according to the BHK interpretation, that we have or will have a proof for a or a proof that a implies a contradiction. Now, in fact, for many mathematical statements we have neither a proof nor a refutation and do not know whether we (i.e. the creative mathematical subject) will ever have a proof or a refutation for these statements. Intuitionism rejects the validity of the law of excluded middle

---

[11] See Troelstra and van Dalen (1988). For readers not familiar with the symbols of formal logic, we recommend reading Chap. 7 first.

for such statements. As an example, consider Goldbach's conjecture: Every even number greater than 2 is the sum of two prime numbers. We have neither a proof of the conjecture nor a refutation and it is uncertain whether a proof of the conjecture or an even number, greater than 2, that is not the sum of two prime numbers, will ever be found. We therefore do not know whether Goldbach's conjecture has a truth value in intuitionistic logic. If it is unprovable, it has no truth value in intuitionism. Kurt Gödel's first incompleteness theorem further states that in any contradiction-free sufficiently strong formalisation of arithmetic there are statements that are neither provable nor refutable. Thus, the law of excluded middle is not valid under any circumstances in intuitionistic logic. Another difference between classical logic and intuitionistic logic is that the double negation of a statement a in intuitionistic logic does not imply the statement a, i.e., $\neg(\neg a) \Rightarrow a$ is not evident in intuitionistic logic. If we have a proof of $\neg(\neg a)$, we cannot transform it into a proof of a. If this were the case, the law of excluded contradiction could also be proven in an intuitionistic logic.

The modus ponens $a \wedge (a \Rightarrow b) \Rightarrow b$ and the law of excluded contradiction $\neg(a \wedge \neg a)$ also hold in intuitionistic logic. This is quite easy to see. If we have a proof for a and this can be transformed into a proof for b, then we have a proof for b, so the modus ponens holds. If we had a proof for a and a proof for $a \Rightarrow \bot$, then with the modus ponens we would have a proof for $\bot$, thus $a \wedge \neg a \Rightarrow \bot$ holds and this is the law of excluded contradiction.

Formal calculi have been described that allow all valid formulas of an intuitionistic logic to be derived from axioms using the modus ponens. An example is the minimal calculus of the Norwegian mathematician Ingebrigt Johansson (1904–1987). This calculus differs from a calculus of classical logic only in that the elimination rule for double negation, i.e. the replacement of $\neg(\neg a)$ by a, is omitted.[12] The standard calculus of intuitionistic logic replaces the elimination rule for double negation with $\bot \Rightarrow a$, i.e., from a contradiction follows any statement.[13] In the minimal calculus, $\bot \Rightarrow a$ does not hold. The concept of intuitionistic logic is not quite clear in terms of which calculus is used. However, the law of excluded middle or the elimination rule for double negation do not hold in any intuitionistic calculus.

For mathematical practice, which consists in providing proofs, or refutations for mathematical statements, the differences described above between classical logic and intuitionistic logic are significant. Since the times of Euclid (third century BC), the proof by contradiction has been a central and frequently used proof technique. We assume the negation of the statement a, which we want to prove, and derive a contradiction from this. Since in classical logic $(\neg a \Rightarrow \bot) \Rightarrow a$ holds, we have thus proved a. In intuitionistic logic, however, only $(\neg a \Rightarrow \bot) \Rightarrow \neg(\neg a))$ holds. A proof by contradiction thus only shows $\neg(\neg a)$, and this does not imply, as stated above, that a is provable. Converting a proof by contradiction of a statement

---

[12] See Johansson (1936).

[13] See Kleen (1991).

into a direct proof that is valid according to intuitionistic logic, proves to be very unpleasant in many cases, or is even impossible. This is particularly true for non-constructive existence proofs, such as the proof of the intermediate value theorem of real analysis.[14] In such existence proofs we assume that a mathematical object does not exist and derive a contradiction from this. Intuitionism does not accept such proofs, but demands the construction of the mathematical object whose existence is to be proven.

Another proof technique not accepted by intuitionism is the proof of contraposition. To prove an implication $a \Rightarrow b$, a proof of the contraposition $\neg b \Rightarrow \neg a$ is sometimes given, when a direct proof of the implication does not succeed. In classical logic, the equivalence

$$(a \Rightarrow b) \Leftrightarrow (\neg b \Rightarrow \neg a).$$

holds. In intuitionistic logic, however, the conclusion

$$(\neg b \Rightarrow \neg a) \Rightarrow (a \Rightarrow b)$$

does not hold. Thus, the proof of an implication by proving the contraposition cannot be used in intuitionistic mathematics, just like the proof by contradiction. In view of intuitionistic logic, peculiarities of intuitionistic mathematics, which Sect. 7.4 deals with, emerge.

8.4 Intuitionistic Mathematics

Intuitionistic mathematics differs from conventional mathematics, which is determined by predicate logic and an axiomatic set theory, not only in terms of the logic used. The understanding of both natural numbers and real numbers in intuitionism arises from the two fundamental acts of the creative subject and does not coincide with the conventional understanding.

In intuitionistic mathematics, there are no actually infinite mathematical objects. An actually infinite mathematical object has infinitely many parts, all of which exist at a given point in time. In particular, an infinite set, such as the set of natural numbers, is such an object. The natural numbers are potentially infinite in intuitionistic mathematics. That is, new numbers can always be added to the natural numbers, but they never exist as a whole, since all their elements cannot be written down or mentally represented at one time. In this sense, intuitionism, like other constructivist theories, is finitist.[15] Although intuitionism denies the existence of the set of natural numbers, it is still able to construct an extensive arithmetic, since

---

[14] See textbooks on analysis for this.

[15] Compare also Chap. 10.

it accepts the principle of complete induction. Brouwer had already dealt with this in his dissertation and developed the following intuitionistic perspective:[16]

Let $a(n)$ be a statement, dependent on a natural number n, and let $a(1)$ be proven. That $a(n)$ implies the statement $a(n + 1)$, means in intuitionism that we have a construction that allows us to transform a proof of $a(n)$ into a proof of $a(n + 1)$. If this is the case, we have a potentially infinite sequence of proofs for the potentially infinite sequence of statements

$$a(1), a(2), a(3), a(4) \dots.$$

Intuitionistically, this means that $a(n)$ holds for all natural numbers n.

Today's intuitionistic arithmetic contains many central results of conventional arithmetic, such as the fundamental theorem, which states that every natural number can be uniquely represented as a product of prime numbers, apart from the interchange of factors. The proof of these results is however, quite cumbersome in intuitionistic mathematics, since complete induction can be used, but no proof by contradiction.

The differences between intuitionistic and conventional analysis and function theory are much more significant than the differences in the field of arithmetic. This is due to the intuitionistic construction of real numbers, which is based on choice sequences. A choice sequence is a potentially infinite sequence $a_0, a_1, a_2, \dots$ of natural numbers, or other constructed mathematical objects, which the creative subject determines successively by free choice. The entries of a choice sequence can be determined by a rule or an algorithm, but they can also be chosen arbitrarily, without following any rule. The creative subject can also follow a rule in the selection up to any entry and then start to choose the entries of the choice sequence without a rule, or it can conversely first choose the entries arbitrarily and from an entry start to follow a rule. In no case is the entry of a choice sequence determined by previous entries. The continuum, i.e., the totality of all real numbers, is determined by all choice sequences $a_0, a_1, a_2, \dots$, where a_0 is a freely chosen natural number and all further entries are the digits of the number, which are freely chosen from $\{0, 1, \dots, 9\}$. If this description of real numbers is taken as a basis, far-reaching consequences arise. It immediately follows that intuitionism cannot distinguish between rational numbers, whose sequence of digits is periodic, and irrational numbers, whose sequence of digits is not periodic, i.e., in intuitionism, $x \in \mathbb{Q} \vee x \notin \mathbb{Q}$ is false for all real numbers x. In intuitionistic mathematics, we cannot even determine whether a real number is zero or non-zero. Finally, in a choice sequence, after any number of entries that are zero, we can choose an entry that is not zero. The piecewise defined function f, given by

$$f(x) = \begin{cases} 1, & x > 0 \\ 0, & x \leq 0 \end{cases},$$

---

[16] See Brouwer (1911) for this.

of real numbers in itself, is not well-defined in intuitionistic mathematics. For the function to be well-defined, we must be able to decide whether a real number x is greater than zero or less than or equal to zero, and this is not possible according to intuitionism. Intuitionistically, the function f is only defined on the set $\{x \mid (x > 0) \vee (x \le 0)\}$ and not on the whole real numbers. In intuitionistic mathematics, there is no real function that is well-defined on the whole real numbers or on an interval and has a jump. In intuitionistic mathematics, therefore, the following theorem holds:

Theorem *All functions $f : \mathbb{R} \to \mathbb{R}$ are continuous.*

In classical analysis, this statement is false. There are even functions in it, such as the Dirichlet function

$$f(x) = \begin{cases} 1, & x \in \mathbb{Q} \\ 0, & x \notin \mathbb{Q} \end{cases},$$

which are nowhere continuous. Of course, this function is not well-defined in intuitionistic mathematics, since we cannot decide whether a real number is rational or not.

8.5 Critique of Intuitionism

Since our critique of intuitionism is harsh and sometimes polemical, we first feel it appropriate to appreciate Brouwer's life's work. Brouwer almost single-handedly and against considerable resistance created an independent metaphysical foundation of mathematics and designed a new logic that differs significantly from classical logic in its principles. In addition to his great professional mathematical achievements, we also respect that Brouwer stood up to the influential formalist school of Hilbert in the foundational dispute. Brouwer can also be considered one of the fathers of constructive mathematics, which today forms an active field of research. In the philosophy of mathematics, Brouwer's intuitionism has had a great influence on constructivism, which we will discuss in Chap. 10.

According to Brouwer, the creative subject forms the metaphysical basis of mathematics, since it creates the objects of mathematics. We consider this approach completely unsuitable for constituting a unified foundation of mathematics. We are willing to accept that Brouwer's mind was subject to the two mental processes or mental acts that he identifies as the basis of mathematics. However, these mental processes are almost completely unknown to the author's mind. Perhaps as a child the author did indeed create numbers through the temporal and successive process of counting, but he rather remembers a continuous flow of time. The natural numbers were never represented in the author's mind as a finite sequence that can be continued, but always as a whole. The difference between the subject Brouwer and the author is even greater with regard to the free choice of a sequence of mathematical objects. The author does not know if he has ever arbitrarily chosen a sequence of numbers and he cannot claim that he would be able to do so. Rolling dice does not help here, a random sequence is not a choice sequence. The crucial point is that the creative mathematical subject of Brouwer, the author and probably also the reader may differ considerably. If there are no linguistic misunderstandings, this is indeed the case. The assumption of the unity or universality of the creative mathematical subject or mind, which Brouwer assumes and which his metaphysics of mathematics requires, is untenable. Who or what should decide whether the mental processes of Brouwer, the author or the reader are those of the ideal mathematical subject and thus fundamental to mathematics?

The second problem of intuitionistic metaphysics lies in the assumption that the mathematical subject is permanent and never forgets a proof of a mathematical theorem. This assumption is fundamental to the concept of truth in intuitionism. Mathematical proofs are of course culturally transmitted, but sometimes a proof is forgotten and possibly rediscovered later. The memories of Brouwer's creative subject can therefore not be traced in the cultural transmission of mathematical theorems and proofs. One must rather imagine that all proofs found so far by the creative mathematical subject are contained in a permanent universal memory, which exists beyond human culture. This memory also contains a proof of Goldbach's conjecture, should an early high culture or an intelligent life form on an exoplanet have already found such a proof. Unfortunately, such a memory of mathematical proofs is unknown to us and we are not sure, whether its existence can be coherently imagined at all. On the one hand, the memory must be transcendent, i.e. exist beyond physical space-time. On the other hand, it is constantly changing due to newly added proofs. Be that as it may, intuitionism certainly makes very strong metaphysical assumptions, which one may be sceptical about. In logic, the situation is different, as intuitionism here refrains from making strong assumptions. The bone of contention in the foundational dispute and the essential difference between classical and intuitionistic logic is, as said above, the law of excluded middle, which intuitionistic logic does not accept. Here is a quote from David Hilbert: *To deny this law of excluded middle to the mathematician would be like forbidding the astronomer the use of the telescope or the boxer the use of his fists.*[17] Without endorsing his behaviour in the foundational dispute, we would like to agree with Hilbert here. The law of excluded middle is an extremely useful tool in mathematical practice for proving theorems. There would have to be compelling reasons to persuade us to abandon this tool. This would certainly be the case if a contradiction could be derived from classical logic, which could be avoided by renouncing the law of excluded middle. However, a theorem by Kurt Gödel shows that this is not possible, if classical logic is inconsistent, then so is intuitionistic logic.[18] If the identification of truth and provability, as proposed by intuitionism, were the only possible explanation of the concept of truth of mathematical statements, we would probably also feel compelled to abandon the law of excluded middle. However,

---

[17] See Hilbert (1928).
[18] See Gödel (1933).

this is not the case. According to a correspondence theory of truth, a mathematical statement is true exactly when it corresponds to the mathematical facts. If we base our theory on this, the controversial law of excluded middle obviously applies. A correspondence theory of truth suggests a realistic metaphysics of mathematics, and it may oblige us to assume that there are mathematical facts beyond the cognising subject. Whether these facts are mental, formal, ideal or physical is not significant for a correspondence theory of truth, in this sense it is metaphysically open. Only a strong scepticism about mathematical reality beyond the subject can force us to submit to the yoke of intuitionistic logic, which significantly complicates the proof of some elementary statements of arithmetic.

The fact that intuitionistic mathematics rejects the existence of the set of natural numbers in the sense of the Peano axioms, is strange to us. Developing a concept or mental representation of all natural numbers is considerably easier for us than the mental representation of a long, but finite initial sequence like

$$1, 2, 3, 4, \ldots, 73^{9123} - 1, 73^{9123}$$

of natural numbers. Even the concept of short sequences like $1, 2, 3, 4, 5$ has a certain mental arbitrariness, why does the sequence end at 5 and not at 4 or 3? To put it poetically, every end has an arbitrariness. Since intuitionism starts from the mathematical subject, it should, given our mental disposition, have no problem with the existence of the set of natural numbers. In our view, the natural numbers are perhaps the most elementary and simplest concept of the mathematical subject. We find it easier to understand if a finitism in the philosophy of mathematics is advocated out of concern for the formal consistency of Peano arithmetic. We will go into this in more detail in Chap. 9.

Our discomfort with the intuitionistic construction of real numbers is even greater than our strangeness regarding intuitionistic arithmetic. First of all, we would like to express doubts as to whether the choice sequences of intuitionism can be coherently imagined or determined at all. If a mathematical subject were able to do this, they would certainly not describe the real numbers and probably no other number range either. The pre-theoretical concept of real numbers lies in the continuous line, whose points are real numbers and in which two different points are distinguished as 0 and 1. Any adequate theory of real numbers should, if possible, meet this concept and classical constructions of real numbers such as by Dedekind cuts or Cauchy sequences do this. The intuitionistic theory does not allow to decide whether an arbitrary real number is 0 or on the side with 1 of 0 or on the other side. Since this trichotomy does not apply in the intuitionistic construction of real numbers, it cannot be considered an adequate description of real numbers. From every number range we expect that exactly one number is distinguished as unit 1 and it can be decided, whether a given number is the unit or not the unit. Even this cannot be guaranteed by a construction using choice sequences. We are therefore convinced that the intuitionistic attempt to develop a theory of real numbers must be considered a failure. It is not without a certain irony that it is precisely intuitionism in the philosophy of mathematics that does not do justice to our basic mathematical intuitions.

Formalism

9

Contents

9.1 Formal Systems

Formulas are not only an instrument for formalising mathematical theories, but also the basis of formalism in the philosophy of mathematics. A formula is a finite sequence of primitive symbols from an alphabet. Primitive symbols are well-distinguished and uniquely identifiable signs that can be physically represented, for example, by ink strokes on a piece of paper or bytes in a digital storage medium. In the philosophy of mathematics, symbols and formulas are not considered as abstract objects that are removed from physical spacetime, but as quasi-concrete objects that can be concretely represented. For example, $\square$, § and ♠ are symbols and §§□♠□♠ is a formula. Depending on which edition of this book the reader has, this is a representation of the symbols by printer's ink on paper or pixels on a screen.

A formal language consists of an alphabet of primitive symbols and a grammar that determines which formulas are well-formed and thus permissible. The most comprehensive grammar allows all finite sequences of symbols to be well-formed formulas. A formal system consists of predefined formulas of a formal language, the axioms and rules, which allow further formulas to be formed from given formulas; these are the inference rules of the system. Inference rules are relations that determine in which case a well-formed formula A can be derived from given well-formed formulas $B_1, \ldots, B_n$; we usually write $B_1, \ldots, B_n \vdash A$ for this.

© The Author(s), under exclusive license to Springer-Verlag GmbH, DE, part of Springer Nature 2025
J. Neunhäuserer, *Introduction to the Philosophy of Mathematics*, Mathematics Study Resources 22, https://doi.org/10.1007/978-3-662-72179-7_9

A well-formed formula A is derivable from the axioms in a given formal system $\mathfrak{F}$ if and only if the formula can be formed by a finite sequence of applications of the inference rules from the axioms; we write $\mathfrak{F} \vdash A$ for this. Examples of formal systems include calculi of classical logic or intuitionistic logic.[1] Formal systems and especially logical calculi are purely syntactic in themselves, their formulas have no meaning. Formulas do not represent statements within a formal system, they do not refer to the world beyond the formal system and have no truth value. One can only speak of a formula being derivable or not derivable. We can understand a formal system as a game with symbols according to fixed rules. In this analogy, a formula is a game state and derivable formulas are the game states that can occur during the game. Formal systems can in principle be implemented in a suitable programming language on a computer. We first store a digital representation of the axioms and write a program that applies all the inference rules to a stored set of formulas and stores the resulting formulas, together with the formulas already generated. Of particular importance are decidable formal systems. For these, there is an algorithm that decides in a finite number of steps whether a given formula is well-formed and derivable. With this, we can determine the status of a formula in a decidable formal system in finite (but possibly very long) time by a computer program.[2]

The formalisation of axiomatic mathematical theories through formal systems is an important tool of modern mathematics. The procedure is as follows: First, a sufficiently strong formal language is specified, in which the axioms of a mathematical theory can be formulated. This typically consists of symbols for constants, variables and certain relations as well as logical symbols. Then the axioms and definitions of the mathematical theory are represented by formulas of this language. Furthermore, suitable inference rules are given, which allow the formulas of the theory to be derived from the axioms. Mostly the inference rules of classical predicate logic are used, but sometimes other logical calculi, such as that of intuitionistic logic, are used. In our view, the formalisation of a mathematical theory is successful if correctly derived formulas can be interpreted as statements, which are true if the axioms are true. Successful formalisations exist, for example, for number theory, determined by the Peano axioms, mathematical analysis, determined by the Hilbert-Bernay formalism or set theory in the sense of the axiom system of Zermelo and Fraenkel.[3]

The benefit of formalising a mathematical theory is that the assumptions of the theory can be clearly identified, which can eliminate misunderstandings that arise in the communication of the theory. If the proofs of a theory are formalised, they can be checked step by step to detect any fallacies. Formalised proofs of theorems of a mathematical theory can also be checked by using computer programs that

---

[1] Compare the presentations in Chaps. 7 and 8.

[2] We give examples of decidable formal systems in connection with Hilbert's programme in Sect. 9.3. Further information on formal systems can be found, for example, in Kleen (1991).

[3] See Sect. 7.4 on Peano arithmetic, Hilbert and Bernays (1934/1939) on the formalism of analysis and the appendix (Chap. 14) for an introduction to axiomatic set theory.

represent the formal system of the theory. If we have an effective formalisation of a mathematical theory, it is fundamentally even possible to find proofs for conjectures using computer programs. Currently, in most cases, the effectiveness of the algorithms used or the available computing power is not sufficient for this. Perhaps one day quantum computers will find proofs for mathematical conjectures that have not yet been found by humans. An efficient proof machine, on which formal systems are implemented, would be a useful tool, as it is often time-consuming and nerve-wracking to track down a proof for a mathematical conjecture. Even if such a machine is built, the task of defining useful or beautiful mathematical objects and formulating conjectures about their properties would still remain for a researching mathematician. The pure mathematician is concerned with the beauty of mathematical objects and the applied mathematician in interdisciplinary collaboration about their usefulness. The learning and teaching mathematician would still have the task of understanding proofs, preparing them in a pleasing way and conveying his understanding to students. We are convinced that no machine can take over these tasks.

9.2   The Formalist Position

That many mathematical theories can be successfully formalised and that such formalisations are useful, seems to be uncontroversial today. The formalist position in the philosophy of mathematics, however, goes much further. Roughly speaking, formalism claims:

Mathematics is the science of formal systems, all of mathematics is formalisable.

The thesis can be interpreted in different ways. A radical formalism, which in the English-speaking literature under the term *game formalism*, asserts that there are no mathematical statements that have a meaning and to which a truth value can be attributed, i.e., that can be true or false. The sentences of mathematics are formulas and nothing more than formulas, which result from playing with formal systems by applying rules. The concepts of mathematics such as sets, relations, functions, numbers, etc. do not refer to objects, but their meaning consists solely in their use in a formal system. A radical formalism can be found, for example, in Wittgenstein, Carnap and Quine, we will go into more detail on this in Sect. 8.3.

A modern formalism agrees with a radical formalism in that the sentences of mathematics within formal systems are meaningless and have no truth value. However, it also assumes that statements of the form

The sentence, represented by the formula A, can be derived or not derived in the formal system $\mathfrak{F}$,

are part of mathematics. The formula $\mathfrak{F} \vdash A$ thus represents a mathematical statement and depending on whether there is a derivation of the formula A in the formal system $\mathfrak{F}$ or not, the statement is true or false. Consider, for example, Bertrand's postulate:

For every natural number, there is a prime number that is greater than this number, but less than or equal to twice this number.

In the usual contemporary logical-set-theoretical formalism, this sentence is formalised as follows:

$$\forall n \in \mathbb{N} \; \exists p \in \mathbb{P} \; : \; n < p \leq 2n \qquad (A).$$

According to the formalistic philosophy of mathematics, Bertrand's postulate or the above formula has no meaning and no truth value. However, consider the statement:

In the axiomatic set theory of Zermelo and Fraenkel, together with the calculus of a classical predicate logic ($3\mathfrak{F}$), the formula A can be derived; formally $3\mathfrak{F} \vdash A$,

according to a modern formalism, this has the truth value true, i.e., a corresponding derivation is known. This approach in the philosophy of mathematics is also referred to as formalism with a meta-mathematical theory or with a proof theory. Such a formalistic position was first consistently developed by the American mathematician and logician Haskell Curry (1900–1982).[4] Like radical formalism, a modern formalism in the sense of Curry also rejects the question of what the sentences of mathematical theories, such as set theory, number theory or analysis, refer to. All these theories manipulate formulas and nothing more. The statements of the metatheory, which exists for every mathematical theory, however, refer to facts, namely to the derivability of a formula in a formal system.

In the philosophy of mathematics, in addition to radical formalism and its extension by a proof theory, there are also differentiated variants. These usually claim that the concepts and sentences of higher mathematics, such as analysis, do not refer to objects, but are only a formal tool. The sentences of elementary mathematics, such as arithmetic, however, are supposed to be substantial and refer to formal objects. Where the boundary between higher and elementary mathematics is drawn in this context is not clearly defined. The most prominent representative of a sophisticated differentiated formalism was undoubtedly David Hilbert (1862–1943). For Hilbert, finite mathematical objects are concrete and the sentences of finite mathematics refer to these. Infinite mathematical objects, however, do not exist and the concepts and sentences of transfinite mathematics are understood by Hilbert in the sense of a radically formalistic philosophy. We will discuss Hilbert's formalistic programme and the problem of the consistency of formal systems, which is of great importance for Hilbert's philosophy, in Sect. 9.4.

---

[4] See Curry (1941).

9.3   Radical Formalism

In the nineteenth century, radical formalistic positions were represented more by mathematicians such as Eduard Heine (1821–1881) and Johannes Thomae (1840–1921) than by philosophers. The philosopher and logician Gottlob Frege (1848–1925) deals with these approaches in his work *Basic Laws of Arithmetic (1903)* and finds that they are neither a unified nor a consistent position in the philosophy of mathematics.[5]

In his work *Tractatus logico-philosophicus 1921*, the Austrian philosopher Ludwig Wittgenstein (1889–1951), among other things, attempts to develop a formalistic philosophy of mathematics.[6] According to Wittgenstein, mathematical sentences have no truth value, only statements that are contingent, i.e. not necessary, should have a truth value. For Wittgenstein, mathematics is a calculus that does not express thoughts about the world as it is, but is only a formal instrument. In the Tractatus, Wittgenstein tries to show that the basis of arithmetic is the forms of use of natural language. He introduced operations as mappings of statements in a non-mathematical language to statements in this language and understands the natural numbers as exponents of these operations. The exponents of the operation here describe the repetition of the operation or the mapping.[7] Wittgenstein's formalism allows at best the definition of a very rudimentary arithmetic. Even the definition of the addition and multiplication of natural numbers implicitly assumes the principle of complete Induction, or recursive definition, is presupposed. This principle is neither formalised nor discussed in the Tractatus. In Wittgenstein's later work, there is the assertion that everything in mathematics is algorithm and nothing is meaning.[8] This thesis can be understood as the credo of a radical formalism in the philosophy of mathematics. However, Wittgenstein's attempt to formalise arithmetic is untenable.

The influence of Wittgenstein's work on the philosophy of the twentieth century is remarkably large. First of all, the reference of modern philosophy of language to Wittgenstein should be mentioned. Some authors even speak of a linguistic turn in modern philosophy, which can be traced back to the work of Wittgenstein.[9] Of greater importance for the philosophy of mathematics is logical positivism, which also refers to Wittgenstein. The philosophy of logical positivism was significantly shaped by the Vienna Circle, a group of philosophers, logicians and other scientists who regularly met at the University of Vienna from 1924 to 1936. Prominent representatives of logical positivism were Rudolf Carnap (1891–1970), Herbert Feigl (1902–1988), Moritz Schlick (1882–1936), Hans Reichenbach (1891–1953)

---

[5] See Frege (1893).

[6] See Wittgenstein (2003).

[7] Here points 6.02 to 6.241 in Wittgenstein (2003) are relevant.

[8] See Wittgenstein (1975).

[9] The term linguistic turn is said to go back to the Austrian philosopher Gustav Bergmann (1906–1987), who was also a member of the Vienna Circle.

and many others. At first glance, the philosophy of mathematics in logical positivism does not appear formalistic. Mathematical sentences are described as analytic, i.e., true due to the meaning of the terms used. But if we consider Carnap's principle of tolerance in logic, this impression changes.[10] According to Carnap, there is no morality in logic, everyone is free to design and use their own logic, i.e., their own formal system, just as they please. This position is clearly formalistic. Every formal calculus is part of mathematics according to Carnap, even if it is inconsistent and leads to contradictions. By downplaying or ignoring the meaning of mathematical concepts in logical positivism, the ontological question, what these concepts refer to, does not arise for a positivist. This is a characteristic of the radical departure of positivism from any metaphysical position. The only criterion for the quality of a mathematical theory in positivism is its practical usefulness. Here the question arises how formulas without meaning can be applied in empirical sciences to prove their usefulness. Carnap seems to be aware of this problem. He suggests designing a structure that combines the formalism of logical-mathematical calculi with a logicist system, as designed by Frege, to meet the demands of formalism and logicism equally.[11] However, Carnap fails to describe such a structure, which does not surprise us. It seems at least difficult, if not impossible, to reconcile the logicist basic assumption that mathematical sentences are logical truths, with the formalistic principle of tolerance in logic. The exact positivist position in the philosophy of mathematics remains, in our view, indeterminate between a formalism and a logicism.

One of the most productive and influential American philosophers of the twentieth century was Willard Van Orman Quine (1908–2000), who was in close contact with the logical positivists, but clearly turned against their philosophy. Quine criticised the concept of analytic truth, which is fundamental to logical positivism. He considers the distinction between sentences that are true due to the meaning of the expressions used, and sentences that make a claim about the world, to be an untenable dogma. According to Quine, no sentence is immune to empirical refutation. In his early work, Quine radicalises formalism in the philosophy of mathematics, by avoiding any reference to logical-analytic truths. At this time, Quine advocates a nominalism, i.e., he denies the existence of all abstract objects.

The article *Steps towards a constructive nominalism (1947)* by Quine and the American philosopher Nelson Goodman (1906–1998) can be understood as a manifesto of radical formalism.[12] The sentences of mathematics are supposed to be nothing more than strings of symbols, which make no statement and to which no truth value is assigned. Goodman and Quine compare the sentences of mathematics with the beads of an abacus, which can serve as a calculation aid, but for which the question of meaning and truth does not arise. Symbols are represented by concrete physical objects according to Goodman and Quine, but they are such

---

[10] See *Logical Syntax of Language* in Carnap (1934).

[11] See also Chap. 7.

[12] See Goodman and Quine (1947).

objects! From this it follows that for Quine there are only finitely many symbols and formulas formed from them. Mathematical rules are purely syntactic in nature, they specify how new strings of symbols can be obtained from concrete physical strings of symbols. Considering the analogy between mathematics and an abacus, the possibilities of moving the beads of the abacus correspond to the syntactic rules.

It is noteworthy that Quine distances himself from nominalism and radical formalism in his later work and develops a naturalism with respect to abstract objects, see Chap. 12.

9.4 Hilbert's Formalism

Essential impulses for the formalisation of mathematics came from the German mathematician David Hilbert (1862–1943) and his assistants Paul Bernay (1888–1977) and Gerhard Gentzen (1909–1945) at the University of Göttingen. Hilbert developed a nuanced formalistic philosophy of mathematics, which is based on the distinction between finite and transfinite mathematics. Broadly speaking, finite mathematics deals only with finite mathematical objects, while transfinite mathematics deals with infinite ones. According to Hilbert, the statements of finite mathematics are meaningful and not meaningless, as they refer to symbols. Symbols, for Hilbert, are immediately intuitively given discrete objects that exist before any thought and thus beyond logic. In Hilbert's philosophy of mathematics, symbols are not physical objects, as we can identify them unambiguously independent of space and time. However, they are also not mental constructions, as their properties are objective and not constituted by the cognising subject. Lastly, Hilbert would probably also reject the idea that symbols are Platonic ideas, as symbols are not abstract, but concrete. We get the impression that Hilbert assumes the existence of an ontologically independent class of formal objects that are not reducible to other objects.

Transfinite mathematics is understood by Hilbert in the sense of a radically formalistic philosophy. Its sentences are nothing more than formulas and the proofs of these sentences are derivations of formulas from formulas according to clearly defined rules. In relation to transfinite mathematics, Hilbert also develops an instrumentalist perspective. Transfinite mathematics serves the purpose of providing conceptual and simple proofs for theorems of real finite mathematics.[13]

Which mathematical concepts and methods are part of finite mathematics is not clearly determined by Hilbert and is still being discussed today. It is mostly assumed that primitive recursive arithmetic (PRA) is part of finite mathematics.[14] However, it is controversial whether the PRA fully describes finite mathematics. The PRA represents a formalisation of the natural numbers, which, unlike Peano arithmetic,

---

[13] Hilbert's philosophical thoughts can be found in particular in Hilbert (1922, 1928, 1931), see also Hilbert (1935).

[14] See Curry (1941).

only uses classical propositional logic and no predicate logic, i.e., does without quantifiers.[15] The axioms of the PRA are the tautologies of propositional logic and the description of identity as an equivalence relation. In addition, there are axioms that describe primitive recursive functions and allow the substitution of variables. In the PRA, there is a rule that formalises the principle of complete induction without the use of quantifiers. In the PRA, we can replace the formula $\phi(0) \wedge (\phi(x) \Rightarrow \phi(N(x))$ with $\phi(y)$, where ϕ is any formula with one variable.

The goal of Hilbert's formalistic programme developed in the 1920s was to formalise all mathematical theories and to prove with finite methods that these formalisations are complete, consistent, and decidable. This means that all true sentences of the formalised theory are provable, i.e., formally derivable from the axioms, that no sentences can be derived from the axioms that contradict each other, and that there is a procedure to determine whether a sentence is derivable. Furthermore, Hilbert wanted to show that every sentence of real finite mathematics that can be proven with transfinite methods can also be proven with finite methods. The problem of proving the consistency of number theory with finite methods was already presented by Hilbert in 1900 at the International Congress of Mathematicians, as one of the 23 greatest problems of mathematics.

Hilbert's programme can today be considered failed in its general form. Gödel's first incompleteness theorem shows that any formal system strong enough to formalise elementary arithmetic is either inconsistent or incomplete. Gödel's second incompleteness theorem shows that such a formal system cannot prove its own consistency.[16] From the incompleteness theorems, it follows that the completeness and consistency of Peano arithmetic cannot be proven with methods of finite mathematics, in the sense of a primitive recursive arithmetic. The same also applies to analysis, determined by the Hilbert-Bernays formalism, and set theory in the sense of the axiom system of Zermelo and Fraenkel. Furthermore, the logicians Alan Turing (1912–1954) and Alonzo Church (1903–1995) independently showed in the 1930s that there is no algorithm that determines for every sentence of number theory in a finite number of steps whether it is derivable from the Peano axioms or not.[17] Finally, it should also be noted that we now know sentences of elementary number theory that can be proven using transfinite ordinal numbers, but are not provable in Peano arithmetic.[18]

What remains of Hilbert's programme is Gentzen's proof of the consistency of Peano arithmetic. This proof is based on a primitive recursive arithmetic, extended by the principle of transfinite induction.[19] We are surprised to find that the proof is accepted by some as finitist, even though it uses transfinite ordinal numbers, which in our view are not part of finite mathematics. Further progress in the spirit of

---

[15] Compare also Chap. 7.

[16] See Gödel (1940).

[17] See Church (1936), Turing (1936).

[18] See Chapter 1 in Neunhäuserer (2015).

[19] See Gentzen (1936).

Hilbert's programme can be found in relation to the decidability of mathematical theories. It has been shown that not all mathematical theories are undecidable, Algorithms have been found for some theories that determine whether a sentence is derivable or not. These theories include arithmetic without multiplication (Presburger arithmetic), analytical, Euclidean and hyperbolic geometry, the theory of commutative groups, and a restricted set theory.[20]

In summary, we note that Hilbert was a realist with respect to finite mathematics, the subject matter of which consists of sequences of symbols and is therefore purely formal in nature. Hilbert takes the consistency of the real finite part of mathematics for granted. With respect to transfinite mathematics, Hilbert adopts a radically formalistic stance. Since transfinite mathematics does not refer to objects, Hilbert is concerned about the consistency and not the truth of transfinite mathematics. We know today that this concern cannot be dispelled with the methods of finite mathematics.

9.5   Critique of Formalism

Although we recognise the value of formalising modern mathematics, we believe that a formalistic philosophy of mathematics is misguided. Mathematical theories consist of statements; axioms are statements, definitions are statements, and mathematical theorems are statements. Statements have a meaning and can be assigned a truth value.[21] The philosophical question of what the meaning of mathematical statements consists in, what they refer to, and in what sense mathematical statements are true or false, is certainly difficult to answer; but ignoring this question is not a reasonable option in the philosophy of mathematics. Since our perspective on formalism in philosophy is that of a researching mathematician, we would like to start by appealing to the mathematical intuition of the reader with some mathematical statements.

(1) For every set, there is a power set that contains the subsets of the set as its elements.
(2) Two sets have the same cardinality if there is a one-to-one correspondence between their elements.
(3) A non-empty set and its power set are not of the same cardinality.

(1) is an axiom of set theory, (2) is a set-theoretical definition, and (3) is an important result of set theory about cardinal numbers.

---

[20] See Monk (1976) for this.

[21] We believe that every definition is trivially analytically true, but this is not the crucial point here.

An axiom of number theory, a central number-theoretical definition, and a major number-theoretical conjecture are given by the following statements:

(1) Every natural number has exactly one succeeding natural number.
(2) A prime number is a natural number that has no proper divisors.
(3) Every even number greater than two is the sum of two prime numbers.

The last statement is Goldbach's conjecture, of which we do not know whether it is true or false, as we know neither a proof nor a refutation. Finally, three statements of analysis are given:

(1) Every non-empty set of real numbers that is bounded above has a smallest upper bound.
(2) A function from the real numbers to the real numbers is continuous if the preimage of every open set is open.
(3) Every continuous function from the real numbers to the real numbers that takes values less than and greater than zero has a root.

(1) can be used as an axiom of analysis, (2) is one way to define continuity, and (3) is an important theorem of analysis.

All these statements can be formalised in suitable systems, yet the claim that one of the statements is nothing more than a meaningless formula beyond a formal game, or can be reduced to such, seems absurd to us. We would like to substantiate this impression with three arguments based on mathematical practice. Our first argument is:

(1) In practice, only logical inference rules are used in the construction of mathematical theories.
(2) If mathematical axioms, definitions, and theorems were not statements, then (1) would be irrational.
(3) (1) is not irrational.

It follows that mathematical axioms, definitions, and theorems are statements.

If one looks at the construction of mathematical theories in the relevant literature, (1) is obvious. The only reasonable reason for the exclusive application of truth-preserving inference rules is that the things to which these rules are applied can be true, hence (2). The formalist is left with only the option of disputing (3). To attribute systematic irrationality to mathematicians in their professional practice, by limiting themselves to logical inference rules, seems unacceptable to us. Our second argument takes the following form:

(1) In mathematical practice, only a small set of axioms is selected from the large set of formally possible axioms.
(2) The best (and perhaps only) explanation for this selection is the assumption that the selected axioms are true statements.

From (1) and (2), it follows by an inference to the best explanation that the axioms of mathematics are statements. Assuming the use of logical inference rules, the theorems of mathematics are also statements.

The first premise of the argument is obvious. Since there are a tremendous number of possible axioms that cannot all be taken into account, it is necessary to select a certain small set of axioms from which mathematical theories are built. A formalist will probably not dispute this and would rather be inclined to question the second premise. He could either claim that the selection of axioms is arbitrary, which seems completely absurd, or he would have to offer another explanation as to why certain axioms are selected. We are not aware of any such explanation and are therefore convinced that (2) is the case.

Our third argument against formalism is:

(1) A theory can only be applied if it makes statements.
(2) Mathematical theories are applied.

From (1) and (2), it follows that mathematical theories (that are applied) make statements.

Although this argument seems compelling, it does have a weakness. Even any non-formalistic philosophy of mathematics struggles to explain how and why the statements of mathematical theories are applied. Nevertheless, we hope that all three arguments taken together will convince the reader, that mathematics contains statements and a radical formalism in the philosophy of mathematics is unacceptable.

Let us now consider a differentiated formalism, which claims that a special mathematical theory consists only of formulas without any statement content. The first and second arguments presented above can also be applied in this case. The third argument, however, is not valid for every mathematical theory, as there are mathematical theories that have not yet found any application and perhaps will never find one. In Hilbert's differentiated formalism, the entire transfinite mathematics is just a formal tool and its sentences do not make any statements. However, analysis is a central part of transfinite mathematics and results of analysis are frequently used in empirical sciences. For example, processes in nature are often quantitatively described by the solutions of differential equations and theorems of analysis guarantee the existence and uniqueness of the solutions of these equations and determine their properties.[22] Mathematical existence statements are thus the basis of empirical modelling, which would not be possible without these statements. If it is not known whether solutions to equations exist at all, these cannot obviously

---

[22] For ordinary differential equations, these are the theorems of Peano and Picard-Lindelöf, see Heuser (2009). Proving the existence of solutions to some partial differential equations, such as the Navier-Stokes equation, is one of the biggest problems in contemporary analysis, see Basieux (2004). The fact that these equations are used to describe natural processes seems problematic to us since we do not know if suitable solutions to the equations actually exist.

be used to describe natural processes. If the existence statements for differential equations were just meaningless formulas, containing an existence quantifier, then these solutions could not describe real phenomena and thus find an application in empirical sciences. Other central theorems of analysis that are widely used are fixed point theorems. These theorems guarantee the existence of a fixed point of a mapping with suitable properties on a space with suitable structure. Fixed point theorems, such as Kakutani's theorem, are the basis of equilibrium theories in economics.[23] Again, a formalistic interpretation of these theorems does not do justice to their application as justification for the existence of real equilibrium states. The diverse applications of analysis seem to be a sufficient reason to reject formalism with respect to this part of mathematics. It is quite irritating that Hilbert, who was significantly involved in the further development of analysis and its application, took a formalistic stance with respect to analysis.

Despite all admiration for Hilbert's mathematical achievements, his unification of realism and formalism with respect to finite mathematics seems doubtful. The statements of finite mathematics are supposed to refer to symbols that exist independently of mental processes. However, a thing, whatever its nature, is not a symbol in itself, but a thing only becomes a symbol, when it is understood as such. Symbols and formulas, which consist of symbols, are therefore not independent of mental processes. In particular, number symbols are a cultural construct that depends heavily on mental processes such as constructions and interpretations. It is well known that the symbol systems of numbers differ significantly in different cultures and epochs. A realism with respect to numbers therefore seems much more plausible than the realism with respect to number symbols that Hilbert seems to represent. A realism with respect to numbers, however, contradicts the restriction of realism to the finite part of mathematics. Hilbert and some others shy away from the assumption of the objective existence of actually infinite objects, such as the natural numbers or the real numbers. This assumption would solve the problem of the consistency of transfinite mathematics, which greatly concerned Hilbert. A Platonic realist gives the following answer to this problem: If an axiom system correctly describes a real existing object area, no contradictions will follow from the axioms. Should a contradiction arise, which we cannot exclude, this shows that our axiom system does not correctly describe an object area, and we must choose other axioms. For example, if Peano arithmetic correctly describes the natural numbers, in the sense of a Platonic realism, which we assume, we will not find any contradictions in Peano arithmetic. If we do find such a contradiction, this means that we do not yet understand the natural numbers correctly and need a new axiom system. This cannot be ruled out and would be as regrettable as it is fascinating.

---

[23] See Church (1936).

Constructivism 10

Contents

10.1   The Constructivist Position

A constructivism in the philosophy of mathematics is based on the following ontological basic assumption:

A mathematical object exists if and only if it can be constructed.

According to this ontological thesis, a mathematical existence statement is true exactly when there is a method to determine the object, whose existence is claimed. In classical mathematics, there are many non-constructive existence proofs. A nice example of such a proof shows that there are irrational numbers a and b such that a^b is a rational number. Either $\sqrt{2}^{\sqrt{2}}$ is rational or irrational. In the first case, with $a = b = \sqrt{2}$ we have already found a and b such that a^b is rational. In the second case, we set $a = \sqrt{2}^{\sqrt{2}}$ and $b = \sqrt{2}$ and find that $a^b = \sqrt{2}^2 = 2$ is rational. However, a constructive proof of the statement can also be given. It holds that $e^{\ln(2)} = 2$ and it can be shown, that both the Euler number e and the natural logarithm $\ln(2)$ are irrational. However, there are also mathematical statements that can be proven

J. Neunhäuserer, *Introduction to the Philosophy of Mathematics*, Mathematics Study Resources 22, https://doi.org/10.1007/978-3-662-72179-7_10

in the classical sense, but not constructively.[1] Constructivism in the philosophy of mathematics rejects non-constructive existence proofs. Such proofs fundamentally rely on the law of excluded middle of classical propositional logic: For a statement a, either a or the negation of a is true, i.e., $a \vee \neg a$ is a tautology.[2] Therefore, constructivism must reject the law of excluded middle and base mathematical proofs on an intuitionistic logic that dispenses with this law. We have already introduced intuitionistic logic in Sect. 8.3 and we will see in Sect. 10.2 that intuitionistic logic is indeed fundamental to the development of constructive mathematics.

To define the constructivist position more precisely, we need to explain the concept of construction, in the sense of a method for determining a mathematical object. As stated in Sect. 8.2, intuitionism attributes two acts of the creative subject to mathematics. If we are willing to accept that these acts represent a method for determining mathematical objects, then intuitionism is a constructivist philosophy of mathematics. This view is widely held in the literature. However, for systematic reasons, we prefer to distinguish constructivism in the philosophy of mathematics from intuitionism. In our sense, mental acts do not constitute constructions; a construction of a mathematical object is rather given by an algorithm, i.e., a computational instruction that determines the object. Assuming this interpretation of the concept of construction, constructivism in the philosophy of mathematics claims that a mathematical object exists exactly when it is computable. The concept of computability can be made precise by concepts of theoretical computer science if we base it on the Church-Turing thesis.[3] This states that the functions computable in an intuitive sense are precisely the Turing-computable functions. A function f is Turing-computable if there is a Turing machine that determines the function value $f(n)$ for each argument n from the function's domain. A Turing machine is a universal computational model that models the operation of a computer. Most computational models are as powerful as Turing machines, but there are also some weaker models that can compute less than Turing machines. Quantum computers also cannot compute more functions than Turing machines, they just do it much faster than classical computers.

The computability of mathematical objects is usually traced back to computable functions. In this sense, a number is computable if there is a computable function that determines the digits of the number. If we consider computable functions on the natural numbers, it becomes clear that a computable number can have an infinite sequence of digits. It can be shown that all algebraic numbers, i.e., the solutions

---

[1] A nice example can be found in functional analysis. Let l^1 be the space of absolutely summable sequences of real numbers and l^∞ the space of bounded real sequences. The space l^1 can be embedded in the dual space of l^∞, i.e., in the space of linear functionals on l^∞. Elements that are in the dual space of l^∞, but not l^1, exist if one assumes the axiom of choice, but cannot be constructed. See Schechter (1997) for this.

[2] See also Sect. 7.3 for this.

[3] We only give a brief introduction to the theory of computability here and refer to the relevant specialist literature in theoretical computer science, such as Hoffmann (2022), for details.

of algebraic equations with rational coefficients, are computable. For example, $\sqrt{2}$ or the golden ratio $\phi = (\sqrt{5} + 1)/2$ are computable. The constants of analysis such as the Archimedes constant π and Euler's number e are also computable. A number that can be concretely defined but not computed is the Chaitin constant Ω_T, introduced by the American mathematician and philosopher Gregory J. Chaitin (1947–). Ω_T is the sum of the terms $2^{-|p|}$, where p is a program on a Turing machine T, $|p|$ denotes the length of the program in bits, and we sum over all programs that halt after a finite time,

$$\Omega_T = \sum_{p \text{ halts}} 2^{-|p|}.$$

Ω_T can be interpreted as the probability that a Turing machine T with a randomly chosen program halts. The approximate value of the Chaitin constant strongly depends on the specific type of Turing machine T considered. For a certain Turing machine, the value $\Omega = 0.00787499699\ldots$ was determined.[4] The Chaitin constant is not computable because the halting problem is undecidable. The halting problem is the problem of whether a Turing machine T with a given program p produces a result and halts, i.e., whether $|p|$ is finite. Alan Turing showed that there is no Turing machine that decides for all possible programs p on a Turing machine whether the computation terminates.[5] A strict constructivist would probably have to claim that Ω_T does not exist, i.e., is not well-defined in the constructivist sense.

The concept of computability can also be introduced for subsets A of countable sets using computable functions, although in this context we usually speak of the decidability of a set. A set A is computable or decidable if the characteristic function of the set, which has the value 1 for arguments in the set and otherwise the value 0, is computable. For example, the set of all prime numbers is computable, as a Turing machine can decide whether a number is a prime number or not. However, a Turing machine cannot decide whether a Diophantine equation has a solution.[6] The set of all solvable Diophantine equations is therefore not computable. Again, a representative of a strictly constructivist philosophy of mathematics would have to claim that such a set does not exist.

After a brief historical digression, we will go into detail about the weaknesses of constructive mathematics and problems of constructivism in the philosophy of mathematics.

---

[4] See Calude and Dinneen (2007) for this.

[5] See Turing (1936).

[6] A Diophantine equation is given by $f(x_1, \ldots, x_n) = 0$ for a polynomial function f with integer coefficients. We are looking for integer solutions $(x_1, \ldots, x_n)$. An example is Pell's equation $x_1^2 - dx_2^2 - 1 = 0$. The Russian mathematician Yuri Matijassevitsch (1947–) proved that the solvability of Diophantine equations is undecidable, see Matijassewitsch (1993).

10.2 Historical Development

We provide here a brief outline of the development of constructive mathematics with a focus on aspects that appear relevant to constructivism in the philosophy of mathematics. The mathematics of intuitionism can be understood in a broader sense as constructive. As stated in Sect. 7.4, it is based on the construction principle of choice sequences and not on the concept of algorithmic computability. With the fundamental work of Alan Turing (1912–1954) and Alonzo Church (1903–1995) on algorithmic computability in the 1930s, constructively recursive mathematics emerges. This considers recursive functions on subsets of the natural numbers and constructs the real numbers with their help. It has been shown that the class of recursive partial functions coincides with the class of Turing-computable functions, which we introduced in Sect. 9.1.[7] One of the main representatives of constructively recursive mathematics was the Russian mathematician Andrei Andreyevich Markov junior (1903–1979), the son of the famous Russian stochastiker of the same name. Markov describes a class of algorithms that exactly computes the recursive partial functions, and assumes an intuitionistic logic that dispenses with the law of excluded middle.[8] In addition, Markov assumes a principle that can be seen in a certain sense as a replacement for the law of excluded middle: Let an infinite sequence of zeros and ones be given. If we assume that the assumption that all sequence members are zero leads to a contradiction, it follows according to classical logic that an entry of the sequence is one. This conclusion is not valid in intuitionistic logic. Markov now assumes that a Markov algorithm, i.e., a Turing machine, will track down the entry of the sequence that is one. Thus, the statement that an entry of the sequence is one is also true in constructive mathematics. The constructively recursive mathematics with Markov's principle is also referred to as Russian recursive mathematics and appears to be a main current of constructive mathematics. However, Markov's principle is not accepted by all representatives of constructivism.

A disturbing result of constructively recursive mathematics is Specker's theorem from the 1940s:

Theorem *There is an increasing, upper-bounded sequence of rational numbers that does not converge to any real number.*[9]

Such a sequence obviously converges to a real number in classical analysis, see Fig. 10.1. Here, the convergence of the sequence follows directly from the supremum principle, which states that every non-empty, upper-bounded set of real numbers has a smallest upper bound. In fact, the supremum principle in constructive mathematics only applies to order-localised sets of real numbers, the distance to any given real number of which is computable.

---

[7] See Cooper (2004).

[8] See Markov (1954).

[9] See Specker (1949).

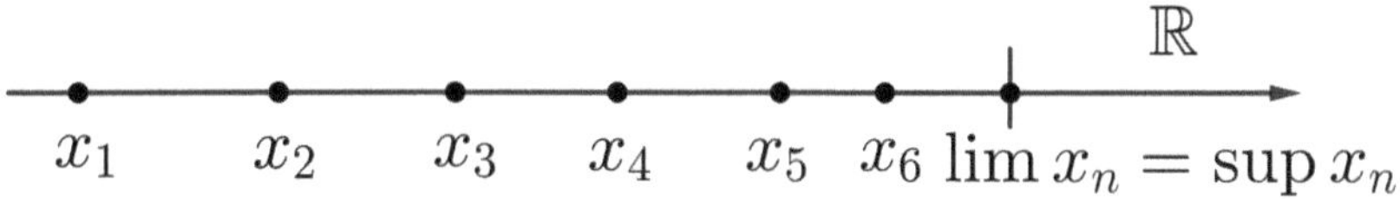

Fig. 10.1 A growing, upper-bounded sequence of real numbers

The book *Foundations of Constructive Analysis (1967)* by the American mathematician Errett Bishop (1928–1983) had a strong influence on the further development of constructive mathematics.[10] In his book, Bishop develops central parts of analysis starting from an intuitionistic logic. He refrains from a formal definition of computability, algorithm, or construction procedure and relies on the fact that in mathematical practice, proofs that use an intuitionistic logic can be considered constructive proofs. From a philosophical perspective, Bishop's renunciation of an ontological statement about which mathematical objects exist is certainly unsatisfactory. However, for the development of constructive mathematics, this open approach was very stimulating. Parts of functional analysis, algebra, and topology could be reconstructed constructively, and modern proof techniques were developed in a constructive form.[11]

Both a constructive type theory and a constructive set theory have been proposed as a formal basis for constructive mathematics. The Swedish mathematician Per Martin-Löf (1942–) developed a type theory that is similar to the type theory of Russell and Whitehead, but is founded on constructivism rather than logicism. According to Martin-Löf, a type is defined by describing what needs to be done to construct an object of that type.[12] In our opinion, a contemporary foundation of constructive mathematics is provided by the axiomatic set theory of the British mathematician Peter Aczel (1941–) and the German mathematician Michael Rathjen.[13] This set theory is based on the system of Zermelo and Fraenkel, which we describe in the appendix (Chap. 14). It replaces the power set, separation, and replacement axioms with weaker constructive principles. Describing the details of constructive set theory is beyond the scope of this book, but we would like to emphasise that the power set of natural numbers, i.e., the set of all sets of natural numbers, does not exist in constructive set theory. The validity of the axiom of choice in constructive mathematics is controversial. In its classical formulation, this principle states that for every family of sets there exists a function that selects an element from each of the sets, see also Sect. 7.5. From the axiom of choice, the law of excluded middle can be derived in a constructive set theory, with the form that this law takes depending on which form of the separation axiom is adopted.[14] In any case, the axiom of choice is not acceptable to a constructivist. On the other

---

[10] See Bishop (1967).

[11] See, for example, Mines et al. (1988) and Picado and Pultr (2011).

[12] See Martin-Löf (1973).

[13] See Aczel and Rathjen (2001).

[14] This is the theorem of Diaconescu-Goodman-Myhill, see van Dalen (1999) and Goodman and Myhill (1978).

hand, in a constructive type theory like that of Martin-Löf, the axiom of choice can be proven, even though the law of excluded middle does not hold, its negation can even be proven. The difference is that in a type theory, sets and their elements are explicitly given, which is not the case in a constructive set theory. The question of the role of the axiom of choice in constructive mathematics seems to us to be not completely resolved to this day.

10.3 Weaknesses of Constructive Mathematics

In this section, we highlight some weaknesses of constructive mathematics and discuss problems of constructivism in the philosophy of mathematics in the next section.

The first problem with constructive mathematics, in our view, is that it is unable to build a theory of infinite cardinal numbers, i.e., the power of sets. In classical mathematics, this is done, starting from an axiomatic set theory, with choice axioms as follows:

Two sets A and B have the same cardinality or power, if there is a bijection, i.e., a one-to-one correspondence between their elements, $|A| = |B|$. The power of A is less than or equal to the power of B, if there is a bijection from B to a subset of A or all of A, $|A| \leq |B|$. The power of A is strictly less than that of B under this condition, if there is no bijection between A and B, $|A| < |B|$. See Fig. 10.2. The Cantor-Bernstein-Schröder theorem now states that if the power of A is less than the power of B and the power of B is less than that of A, a bijection exists between the sets, so they have the same power,

$$|A| \leq |B| \wedge |A| \geq |B| \Rightarrow |A| = |B|.$$

In addition, with the help of the axiom of choice, it can be shown for any two sets A and B that the power of A is less than or equal to the power of B or the power of B is less than or equal to the power of A,

$$|A| \leq |B| \vee |A| \geq |B|.$$

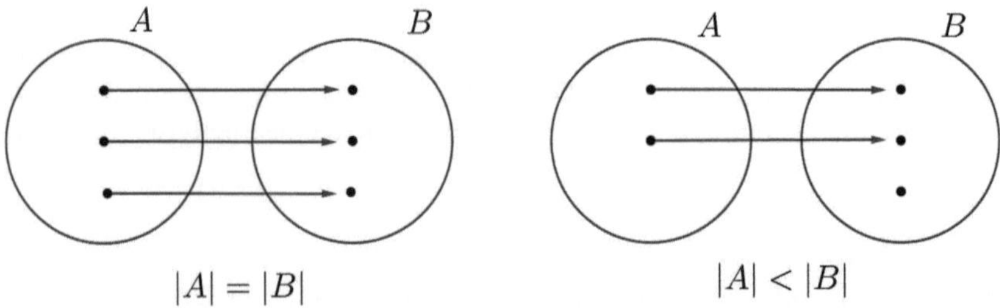

Fig. 10.2 On the definition of cardinality

These theorems are the basis of the classical theory of cardinal numbers.[15] Now the proofs of these theorems are not constructive and there are demonstrably no constructive proofs. The sought-after bijections cannot generally be given explicitly and are generally not Turing-computable.

In the classical theory of cardinal numbers, some wonderful theorems can be proven. The set of Turing machines is countable, there is a bijection between them and the natural numbers. Thus, in classical mathematics, the Turing-computable numbers $\mathbb{B}$ are also countable, i.e., have the same power as the natural numbers $\mathbb{N}$. A simple non-constructive proof by contradiction further shows that the power set $P(M)$ of a set M, i.e., the set of all subsets of M, has a strictly greater power than M. The power of the power set of the natural numbers $P(\mathbb{N})$ coincides with the power of the real numbers $\mathbb{R}$ when these are classically constructed, and is even equal to the power of all sequences of real numbers $\mathbb{R}^{\mathbb{N}}$. The power of the power set of the real numbers $P(\mathbb{R})$ is again greater than the power of $\mathbb{R}$ and coincides with the power of all functions from $\mathbb{R}$ to $\mathbb{R}$.[16] In summary, we have:

Theorem $\mathbb{N} = \mathbb{B} < P(\mathbb{N}) = \mathbb{R} = \mathbb{R}^{\mathbb{N}} < P(\mathbb{R}) = \mathbb{R}^{\mathbb{R}}.$

We regard this statement as a great achievement of modern mathematics, as it explicates the concept of infinity. Regrettably, this statement is not accessible to constructive mathematics. In it, there is no bijection between natural and computable numbers, no power set of natural numbers, and certainly no bijection between this set and the real numbers or all sequences of real numbers. Of course, the power set of real numbers also does not exist in constructive mathematics, nor does a bijection of this set to the set of real functions. The reader will now understand that we cannot muster much sympathy for the restriction of mathematics by constructivism.

The second problem of constructive mathematics, in our view, is that it is not capable of developing a viable measure theory. From a length measure, we naturally expect that the measure of intervals is their length and that the measure of individual numbers is zero. Furthermore, for the development of the theory, we expect a length measure on the real numbers to be an outer measure on the subsets and a measure on the measurable subsets of the real numbers. This means in particular that the measure of a countable union of subsets of the real numbers is less than the sum of the measures of the individual sets,

$$\mu\left(\bigcup_{i=1}^{\infty} A_i\right) \leq \sum_{i=1}^{\infty} \mu(A_i).$$

---

[15] See Chapter 1 in Neunhäuserer (2015).

[16] See again Chapter 1 in Neunhäuserer (2015).

For countable disjoint unions of measurable sets, we expect that the measure of the union matches the sum of the measures of the individual sets,

$$\mu(\bigcup_{i=1}^{\infty} A_i) = \sum_{i=1}^{\infty} \mu(A_i), \text{ when } A_i \text{ is measurable and disjoint.}$$

In classical analysis, the Lebesgue measure meets these requirements.[17] It is easy to see that the Lebesgue measure of the set of Turing-computable real numbers is zero, so this set is negligible with respect to the Lebesgue measure. The definition of the Lebesgue measure and the Lebesgue-measurable sets is not constructive, and we claim that these definitions cannot be reconstructed constructively. Any outer measure on the set of computable numbers is indeed identically zero if we assume that the measure of individual computable numbers is zero. Thus, in constructive mathematics, there is no concept of length that meets our requirements. The same applies to the concept of dimension. We expect a dimension to be countably stable. This means that the dimension of a countable union of sets matches the supremum of the dimensions of the individual sets. In the classical theory, the Hausdorff dimension meets these requirements and cannot be reconstructed constructively. On the set of computable numbers, any countably stable dimension is identically zero if we assume that individual numbers are zero-dimensional. From our perspective, this rules out a constructivist reconstruction.

The author of this book has been able to achieve some results in classical measure and dimension theory in recent years that are not comprehensible in constructive mathematics. The reader will understand that we are not open to the constructivist restriction of mathematics. The list of weaknesses of constructive mathematics could be continued. For example, certain theorems of general topology are not part of constructive mathematics. This is particularly true for all theorems whose proof is based on the existence of ultrafilters.[18]

10.4 Assessment of Constructivist Philosophy

As we have seen in the last two sections, constructive mathematics holds strange results such as Specker's theorem, which contradict classical results. Some classical theories, such as cardinal number and measure theory, cannot be reconstructed constructively. There would have to be good reasons to justify constructivism in the philosophy of mathematics in order to persuade us to engage in the project of constructive mathematics. Not all mathematicians take this perspective, rather a formalist perspective on constructive mathematics is currently prevalent. From this perspective, mathematics can and should examine both constructivist and classical

---

[17] See Bauer (1998) for this.

[18] See Neunhäuserer (2015) or textbooks on general topology for this.

formal systems and determine which theorems can be derived in the classical and constructivist game. The philosophical question of whether only mathematical objects exist that are constructible, or computable, does not arise for a formalist. He can regard contradicting formal systems, in which the existence of non-constructible mathematical objects is assumed or denied, as equivalent objects of mathematical research. Since we reject formalism in the philosophy of mathematics, as explained in the last chapter, we ask ourselves whether non-constructible mathematical objects actually do not exist. In classical mathematics, we have a well-defined concept of real numbers, which successfully formalises the idea of the complete number line. From this concept of real numbers follows the existence of uncountably many non-computable numbers, and we even know examples of non-computable numbers such as Chaitin's constant. This clearly speaks against constructivism in the philosophy of mathematics. Nevertheless, we would like to present and discuss two philosophical arguments that speak for constructivism.

An anti-realist, who has the basic intuition that the objects of mathematics are not independent of mental processes, could put forward the following cogent argument for constructivism:

(1) The objects of mathematics are a product of the human mind.
(2) The human mind is nothing more than a Turing machine in its mathematical capabilities.

From (1) and (2) it follows: Mathematical objects exist exactly when they are Turing-computable.

The first premise of the argument is a way to explicate an anti-realism regarding mathematics. The second premise is a reductionist thesis in the philosophy of mind. Reductionism is a widespread position; mental processes are supposed to be reducible in one way or another to non-mental processes. In the above argument, the possibility of reducing certain mental processes to computational processes, which can be modelled by a Turing machine, is assumed. We believe that this assumption is incorrect and that our mathematical abilities exceed those of Turing machines. We will explain our view using an example. Humanity has had great success in determining whether a given class of Diophantine equations has solutions or not. Our latest major success is the proof of Fermat's Last Theorem by Andrew Wiles (1953–) and Richard Taylor (1962–), which is considered one of the highlights of twentieth century mathematics. This theorem states that equations of the form $a^n + b^n = c^n$ for natural numbers n greater than two have no solutions.[19] To decide whether a class of Diophantine equations has solutions, it is often necessary to develop new terms, theories and proof techniques. We are apparently capable of this. It is therefore reasonable to assume that we are in principle capable of deciding for every class of Diophantine equations whether they are solvable or not. A universal Turing machine is not capable of this and we assume that no other

---

[19] An introduction to the complex proof is provided by Faltings (1995).

universal computing machine, however it may be constructed, can achieve this. Our mental abilities preclude the reduction of all mental processes to non-mental processes. Many readers will certainly not share our view at this point and will argue vigorously against it. However, it cannot be the task of this book to develop a consistent non-reductive philosophy of mind.

A Platonic realist, who has the basic intuition that the objects of mathematics exist independently of mental processes and are abstract, has no good reason to assume that only constructible or computable mathematical objects exist. While we see no logical contradiction between Platonism and constructivism, a Platonist will not want to limit the world of abstract objects to constructible or computable objects and has no reason to do so. The situation is different for a realist regarding mathematics, who wants to see all mathematical objects concretely represented. Such a person could argue as follows:

(1) Mathematical objects exist exactly when they are concretely represented in spatio-temporal reality.
(2) Only computable mathematical objects can be concretely represented in spatio-temporal reality.

From (1) and (2) it follows: Only computable mathematical objects exist.

The first premise of the argument is a way to formulate a non-Platonic realism. We will discuss this position in Chap. 12. The second premise of the argument is a statement about the nature of spatio-temporal reality, i.e., a statement about the physical world in a broader sense. Starting from a finitism regarding the physical world, one could argue for this premise. The spatio-temporal reality is finite, so only finite objects and in particular only finite mathematical objects can be found in it. These objects are computable. We consider this argument to be untenable and the second premise of the argument to be unfounded. Even in a finite universe, infinite and non-computable mathematical objects can be represented. Consider the natural constants, such as the speed of light c, the gravitational constant G or Planck's quantum of action h.[20] These numbers are represented in physical reality. The question of whether these numbers are rational or irrational is open. The rational values of some natural constants, which are set in physics, only approximate the actual values, which can be improved.[21] Rational numbers can be represented by a finite or periodic sequence of digits, for the representation of irrational numbers we need an infinite, non-periodic sequence of digits. Similarly, the question of whether the natural constants are computable in the mathematical sense or not is open. Basically, these questions are not empirical, as there are practical as well as

---

[20] Since 1983, the speed of light in physics has been defined by a constant rational value, but this definition is not immune to revisions. For example, the theory of loop quantum gravity suggests that the speed of a photon cannot be defined as a constant, as its value depends on the photon frequency.

[21] See Mohr et al. (2008).

theoretical limits to the precision of measuring the natural constants. The possibility that the natural constants, like the Chaitin constant, are not computable cannot therefore be empirically ruled out, and it seems at least difficult, if not impossible, to theoretically exclude this possibility. We cannot offer a natural philosophy here that attempts to clarify the difficult question of the number-theoretical properties of the natural constants, but we are convinced that the second premise of the above argument is not justified. Some readers will certainly eloquently disagree.

In this section, we do not intend to discuss all possible arguments for constructivism, but we believe we have highlighted the weaknesses of two essential arguments. We already mentioned in Chap. 8 that we are not willing to submit to the yoke of intuitionistic logic. We have the impression that all attempts to philosophically justify constructivism are not sufficient to compel a researching mathematician to renounce the law of excluded middle, proof by contradiction and parts of classical mathematics. Nevertheless, we can understand that a proponent of a formalistic philosophy finds it appealing to practice constructive mathematics as an equal alternative to classical mathematics.

Structuralism 11

Contents

11.1 Introduction

In this section, we will introduce the basic idea of structuralism in the philosophy of mathematics. To do this, we first need to explain what a structure is. Consider a system that consists of a collection of objects and relations between these objects. A structure is an abstract form of such a system, abstracting from all properties of the objects of the system that are not given by the relations of the system. The objects of a structure are thus completely determined by their relational properties within the structure and they have no intrinsic properties. A system is a realisation or an example of a given structure if the objects of the system exhibit the relational properties of the structure. In this case, we also say that a system exemplifies a structure. Here is an example: A group of people, say a family, together with the kinship relations between the people, forms a system. If we now abstract from the people, the structure of their kinship relations remains. Any group of people or family that exhibits this kinship relationship is a realisation or an example of this structure. As a second example, consider the employees of a company, together with the relation of authority. If we abstract from the employees, the hierarchical staff structure of the company remains. It should be noted here that not for every possible hierarchical staff structure of a company does a company actually exist that realises this. Not every structure is exemplified by a given range of systems.

© The Author(s), under exclusive license to Springer-Verlag GmbH,
DE, part of Springer Nature 2025
J. Neunhäuserer, *Introduction to the Philosophy of Mathematics*,
Mathematics Study Resources 22, https://doi.org/10.1007/978-3-662-72179-7_11

From the perspective of classical metaphysics, a structure is a special form of a universal, i.e., a general concept. Structures are universals that, unlike properties, do not apply to individual objects, but to systems of objects. For example, the universal *human* applies to individual people, the structural universal *nuclear family* applies to a system consisting of parents and children. Whether universals actually exist or whether they are mental constructions is a classic problem of metaphysics.[1]

In contemporary mathematics, structures are described by axiom systems. The axiom system of a structure specifies the relational relationships in which the objects of a mathematical system stand that has the structure. For example, the Peano axioms describe the structure of the natural numbers, the axioms of Euclidean geometry describe the structure of Euclidean space, algebraic axioms describe algebraic structures such as groups, rings, fields, etc. The basic assumption of structuralism in the philosophy of mathematics is now the following:

The subject area of a mathematical theory is given by a structure or several structures that are in relational relationships.

It is not disputed that many mathematical theories investigate structures. It is irrelevant for the statements of these theories how these structures are realised or which examples exemplify the structures. Elementary number theory investigates the structure of the natural numbers and it is not important for its statements whether we realise the structure of the natural numbers using sequences of points, the decimal system or another positional system. Similarly, algebra deals with algebraic structures and a statement about an algebraic structure applies to all systems that represent an example of the structure. It is also not disputed that there are many ways to realise or exemplify each mathematical structure. In particular, there exist many different systems of sets that realise a mathematical structure. In his influential essay *What Numbers Could Not Be (1965)*, the American philosopher Paul Benacerraf (1931–) points out that the structure of the natural numbers can be realised by different set systems.[2] For example, by defining $n + 1 = \{n\}$ and $n + 1 = n \cup \{n\}$, where 0 is the empty set $\emptyset$ in both cases, we obtain two different set-theoretic realisations of the structure of the natural numbers:

1	$\{\emptyset\}$	$\{\emptyset\}$
2	$\{\{\emptyset\}\}$	$\{\emptyset, \{\emptyset\}\}$
3	$\{\{\{\emptyset\}\}\}$	$\{\emptyset, \{\emptyset, \{\emptyset\}\}\}$
4	$\{\{\{\{\emptyset\}\}\}\}$	$\{\emptyset, \{\emptyset, \{\emptyset, \{\emptyset\}\}\}\}$
$\vdots$	$\vdots$	$\vdots$

---

[1] See Stegmüller (1978) for this.

[2] See Benacerraf (1965) for this.

The first realisation of the natural numbers is due to the mathematician Ernst Zermelo (1871–1953) and the second is due to the mathematician John von Neumann (1903–1957). From the point of view of arithmetic, both realisations are equivalent, neither is to be preferred over the other. Benacerraf therefore rules out the identification of a natural number with a set and sees the natural numbers as completely determined by their structure. Benacerraf's observation was an important starting point for the development and discussion of structuralist approaches in the philosophy of mathematics. Whether or in what sense structures exist, how we recognise structures and whether structures are the fundamental or even the only subject of mathematics is disputed. The contemporary literature on structuralism in the philosophy of mathematics is diverse and confusing. We believe we can identify four basic positions. We present these positions in Sects. 11.2–11.5 and discuss them.

11.2 Abstract Structuralism

Abstract structuralism in the philosophy of mathematics is also called Platonic structuralism or ante-rem structuralism. The English translation of the Latin term *ante rem* is *before the things*. The main theses of abstract structuralism can be formulated as follows:

(1) The fundamental objects of mathematics are structures.
(2) Mathematical structures exist independently of mental processes and are abstract; they are not part of the space-time reality.

Abstract structuralism is lively discussed in the philosophy of mathematics. The American philosophers Michael Resnik (1938–) and Stewart Shapiro (1951–) are considered significant advocates of this position.[3]

In abstract structuralism, mathematical structures have ontological priority; they are the objects of mathematics that exist. Each abstract structure is constituted in ante-rem structuralism by places or vacancies and relations of these places, forming a whole from these two things. Consider, for example, the abstract structure of a company. It consists of the positions or posts that can be occupied by employees, and the relations of these positions. Which employees occupy the positions is irrelevant for the abstract structure of the company.

All objects of mathematics that are not structures are defined by places in a structure. These objects are nothing more than vacancies of the relations that determine the structure. All individual objects of mathematics are identified with individual places in a structure and thus possess no intrinsic properties. Singular mathematical terms accordingly denote individual places or vacancies in a structure. Terms such as 0, 1, 2 etc. refer to nothing other than well-defined places in the

---

[3] See Shapiro (1997) and Resnik (1997).

structure of natural numbers. An equation like $3 + 5 = 8$ expresses a truth about this structure. Similarly, singular terms like *point, line, plane* refer to well-defined places in the geometric structure. A statement like *A line that has two points in common with a plane lies in it* is a truth about the structure of a geometry. Mathematical structures in abstract structuralism are free-standing, that is, independent of systems that realise them. Abstract structuralism does not require a background ontology that postulates the existence of systems that realise a structure, as it presupposes the independent existence of structures. Which objects occupy places in a structure or which terms denote such places is arbitrary; the places of the structure exist for themselves. It is meaningless for abstract structuralism whether the places in the structure of natural numbers are occupied by sets, as described by von Neumann, and $\{\{\emptyset\}\}$ plays the structural role of two, or whether the structure of natural numbers is realised by the sets of Zermelo and $\{\emptyset, \{\emptyset\}\}$ plays the role of two. The places in the structure itself are objects of number theory. In general, in abstract structuralism, it is irrelevant which system of sets realises a structure, as the places and the relation of the structure exist for themselves. This is a solution to the problem of Benacerraf, which we described in Sect. 11.1.

Abstract structuralism, like any form of Platonism, is confronted with the metaphysical gap that we have already presented in Sect. 3.4. How do we, as a space-time life form, gain knowledge about structures that are not part of the space-time reality? Shapiro develops an epistemology that attempts to answer this question. The ability of abstraction, projection and description are supposed to allow us to gain knowledge about abstract structures. Abstraction enables us to directly comprehend the abstract structure of small systems. With a projection, we transfer our knowledge of smaller abstract structures to larger ones. Finally, a description, if coherent, also allows us to gain knowledge about structures that are not directly accessible to abstraction and projection. We do not want to discuss Shapiro's epistemology in detail here, but we would like to note that we miss the mathematical intuition as a source of knowledge of abstract structures in Shapiro's approach. In our view, the serious problems of abstract structuralism are different.

First, we note that abstract structuralism fails with respect to all of mathematics, in particular, set theory cannot be understood purely structuralistically. Sets are fundamental objects of mathematics that have both structural and intrinsic properties. A set is uniquely determined by its elements, and the property of a set to contain or not contain an element cannot be reconstructed structuralistically. The property of a set to contain or not contain an element is not given by a relation between sets. In particular, the empty set, which we cannot do without in set theory, is characterised by the fact that it contains no elements, and this property is obviously intrinsic and not structural. A proponent of abstract structuralism in the philosophy of mathematics might be tempted to segregate sets from structural mathematics and claim that the fundamental objects of not all, but most mathematical theories, are abstract structures. We will show below that this view is questionable, as sets seem to be more fundamental objects of mathematics than structures.

Starting from a set theory, as determined by the axiom system of Zermelo and Fraenkel, structures can be formally defined.[4] A relation R on a set A is in set theory a set of pairs (a, b) or tuples $(a_1, \ldots, a_n)$ with elements from A. A pair is the set $\{a, \{a, b\}\}$ and a tuple can also be defined set-theoretically. A system S can now be understood as a pair of the set A and a set of relations R_i on the set A,

$$S = (A, \{R_i \mid i \in I\}).$$

Two systems are isomorphic, i.e., they realise the same structure, if there is a relation-preserving mapping between them.[5] Finally, a structure is defined as a set of isomorphic systems. So, if abstract sets exist, abstract structures also exist as certain types of sets. This definition of a structure contradicts the first basic assumption of abstract structuralism, that structures are fundamental in mathematics. A representative of abstract structuralism should have a concept of structure that does not rely on the concept of sets. In the philosophical literature, the concept of structure is used as a basic concept that is only explained in a circular way and is not well-defined. An axiomatic and non-set-theoretical definition of structures, which is strong enough to serve as the basis of mathematical theories, is unfortunately not found. Only such a definition could convince us that structures in mathematics are as fundamental as sets.[6]

11.3 Concrete Structuralism

The central thesis of the philosophy of mathematics, which we call concrete structuralism, is:

Mathematical structures exist as the structures of concrete systems and only as such.

In the philosophical literature, concrete structuralism is mostly called *In-re-Structuralism*.[7] This means that the mathematical structures are found in things. We believe that the term *In-re-Structuralism* does not clearly identify the core of concrete structuralism. If mathematical structures are understood as the structures of abstract systems, these are also *in re* in a certain sense, but this is not the position of concrete structuralism.

That a mathematical structure exists as the structure of a concrete system means that it is realised or exemplified by at least one concrete, i.e., spatio-temporal

---

[4] We describe this axiom system in the appendix (Chap. 13).

[5] It is quite a technical matter to specify this. Two systems $(A, \{R_i \mid i \in I\})$ and $(\tilde{A}, \{\tilde{R}_j \mid j \in J\})$ are isomorphic if there are bijections $f : A \to \tilde{A}$ and $g : I \to J$ such that: $(a_1, \ldots, a_{n_i}) \in R_i$ if and only if $(f(a_1), \ldots, f(a_{n_i})) \in \tilde{R}_{g(i)}$.

[6] We will return to this topic in Sect. 13.4 on category theory.

[7] This nomenclature can be found in Shapiro (1997) and seems to be generally accepted.

system. The relations that determine a mathematical structure are thus relations between things in spatio-temporal reality. Concrete systems are therefore given ontological priority in concrete structuralism. If no concrete system exists that realises a mathematical structure, this structure does not exist.

Concrete structuralism can be understood as Aristotelian structuralism in the philosophy of mathematics. We had already pointed out in Sect. 11.1 that structures in the terminology of classical metaphysics are universals. Aristotle, in contrast to Plato, believed that the being of individual things has priority over the being of universals. For Aristotle, universals are an abstraction from individual things; they exist, but their existence is not independent of individual things. Universals only exist if individual things also exist, to which they belong and from which they are derived.[8] If we now consider concrete systems of individual things and their structures instead of individual things, we get from the Aristotelian approach the basis of concrete structuralism in the philosophy of mathematics.

The advantage of concrete structuralism over abstract structuralism is that it does without the assumption of a separate class of abstract objects that exist completely independently of the spatio-temporal world. Many modern philosophers find a Platonic ontology, in whatever form, too speculative. Nevertheless, Aristotelian structuralism does not play a major role in contemporary philosophy of mathematics. We see the reason for this in a widely accepted argument that speaks against concrete structuralism.

Some mathematical structures are indeed realised by concrete systems. Mathematically oriented empirical scientists search for and find mathematical structures in spatio-temporal systems. Sometimes the approach is also the other way round. For structures that have been mathematically described and studied, concrete systems are sought and found that realise these. The second approach is particularly found in theoretical physics. However, we do not know concrete systems that realise all the structures that are the subject of mathematical theories. A proponent of concrete structuralism has two ways of dealing with this fact. He can claim that concrete systems exist for all mathematical structures that realise these, some of these systems are simply not yet known to us. Or he can claim that certain structures, which mathematical theories talk about, do not exist. The first alternative is a far-reaching speculation about the mathematical nature of the spatio-temporal world, which can hardly be justified. Moreover, the cardinality of all concrete systems seems to be limited by the cardinality of the entire physical universe, i.e., its size in the set-theoretical sense. However, the cardinality of the mathematical structures we can describe is unlimited.[9] As far as we know, no one claims this even for the set

---

[8] See again Stegmüller (1978).

[9] If we consider the power set $P(M)$ of a set M, i.e., the set of all its subsets, it has a larger cardinality than M. Starting from the natural numbers, $|\mathbb{N}| < |P(\mathbb{N})| < |P(P(\mathbb{N}))| < |P(P(P(\mathbb{N})))| \ldots$ describes an unlimited sequence of cardinalities. See Chapter 1 in Neunhäuserer (2015). Even if the real numbers with the cardinality $|P(\mathbb{N})|$ and also the set of all functions from the real numbers to the real numbers with the cardinality $|P(P(\mathbb{N}))|$ should be concretely realised in the physical universe, it is not to be expected that large cardinalities are concretely realisable.

$P(P(P(\mathbb{N})))$. The second alternative makes the existence of mathematical objects dependent on empirical findings, namely the existence of space-time systems, and thus declares certain mathematical theories to be objectless and meaningless. Such a fundamental splitting of mathematics and discrediting of some mathematical theories based on an empirical criterion seems absurd to us.

11.4 Eliminative Structuralism

Eliminative structuralism in the philosophy of mathematics is also called *post-rem structuralism*. The translation of the Latin term *post rem* is *after the things*. The main theses of this position can be formulated as follows:

(1) A statement about a mathematical structure is nothing more than a statement about systems with given structural properties.
(2) There are no structures as independent objects of mathematics.

Paul Benacerraf (1931–) and Oystein Linnebo (1971–) can be read as representatives of this position in the philosophy of mathematics.[10]

According to the first thesis of eliminative structuralism, talking about mathematical structures is just a way of talking about systems of mathematical objects with certain relations between these objects. The second thesis states that there are no structures with places or vacancies to which mathematical terms refer, as abstract structuralism claims. Eliminative structuralism also rejects the assumption of in-re structures, i.e., universals that apply to mathematical systems. A statement about a mathematical structure in eliminative structuralism is not a statement about an ante-re or in-re structure. Instead, it refers to all mathematical systems that have certain given properties. Such a statement expresses commonalities of individual mathematical systems and is in this sense a generalisation. Consider, for example, the sentence *Every natural number is the product of prime numbers* in arithmetic. This, interpreted eliminatively, means that in every system of natural numbers, every number of the system is a product of prime numbers of the system. The concept of a prime number is determined here by referring back to the relations that define a system of natural numbers.

In order for statements about a mathematical structure not to be objectless and thus meaningless, a representative of eliminative structuralism must assume the existence of mathematical systems that realise these, i.e., contain the required structural relations. He needs a strong background ontology that guarantees the existence of sufficiently many mathematical objects and mathematical systems. For each mathematical theory that deals with structures, there must be abstract objects that prevent the theory from becoming objectless. The ontology of a sufficiently strong axiomatic set theory offers itself as the basis of eliminative structuralism.

---

[10] See Benacerraf (1965) and Linnebo (2008).

Such an ontology asserts the existence of a variety of sets and set systems.[11] We are not aware of any structure, in whatever sense, that could not be realised by a system of sets. As we have already shown in Sect. 10.2, mathematical systems can be defined by sets and relations on these. According to this definition, a statement about a mathematical structure becomes a statement about isomorphic set systems. Sets are thus the ontologically fundamental objects of mathematics. What remains of structuralism in its eliminative and set-theoretically founded form is the following thesis:

The objects of most mathematical theories are structures in the sense of sets of isomorphic set systems.

If one understands the concept of the set and the set system sufficiently generally, this thesis is demonstrably true for almost all mathematical theories. In this sense, mathematics is indeed the science of structures, as is sometimes claimed.

We assume that the variant of eliminative structuralism that we have outlined here is accepted by many mathematicians and some philosophers. Of course, there are also mathematicians and philosophers who criticise the Platonic realism with respect to sets that we presuppose. Since the problems of Platonism were already discussed in Chap. 3, we do not want to go into them again here. Unfortunately, the reduction of the concept of structure to the concept of set is not simple. Therefore, an axiomatic definition of structures that does without sets would certainly be desirable. In the literature on structuralism in the philosophy of mathematics, we have unfortunately not been able to discover such a thing. If such a definition were available, we could test how powerful it is as a basis for mathematical theories.

11.5 Modal Structuralism

Modal structuralism in the philosophy of mathematics is a variation of the eliminative structuralism that we discussed in Sect. 11.4. The idea of modal structuralism is that mathematical theories do not refer to actually existing, but possible mathematical systems. The central thesis is therefore:

A statement about a mathematical structure is nothing more than a statement about all possible systems with given structural properties.

The American philosopher Geoffrey Hellman (1943–) is the main representative of this position in the philosophy of mathematics.[12]

Like a proponent of eliminative structuralism, a representative of modal structuralism assumes that structures do not exist as independent objects of mathematics, and thus distances himself from abstract structuralism. In contrast to eliminative structuralism, which we described in Sect. 11.4, modal structuralism tries to do

---

[11] We refer again to the appendix of the book.

[12] See Hellman (1989).

without the assumption of the existence of abstract systems, i.e., a rich mathematical ontology. A statement about a mathematical structure does not refer to real mathematical systems, but to possible systems. Consider, for example, the sentence *Every natural number is the product of prime numbers* in arithmetic. This, interpreted in a modal structuralist way, means that in every possible system of natural numbers, every number of the system is a product of prime numbers of the system.

To clarify the position of modal structuralism, an explanation of the concept of possibility is necessary. What kind of possibility is meant when we talk about possible mathematical systems? If metaphysical possibility is meant, strong ontological premises creep in. We must assume the existence of sufficiently many possible mathematical worlds to realise mathematical structures. Nothing is gained here compared to the assumption of sufficiently many set systems. If possible mathematical systems mean physically possible systems, modal structuralism is confronted with the same problems as concrete structuralism. We have already addressed these problems in Sect. 11.3. Hellman understands possibility as logical possibility and uses a formal modal logic to formulate his modal structuralism. When we talk about possible mathematical systems, we mean logically possible systems. The concept of logical possibility also requires explanation in our view. It is common today to understand logical possibility in set-theoretical terms. To assert that a sentence is logically possible means to assert that there is a set that satisfies it. This option is not in line with Hellman, as it again requires a set-theoretical ontology. Hellman seems to use the concept of logical possibility in his work as a basic concept that is explicated by his system as a whole. As a basis for arithmetic, his system includes the assumption that it is logically possible for a countably infinite sequence, and thus a system that realises the structure of the natural numbers, to exist. This assumption corresponds to the assumption of the existence of an inductive set in an axiomatic set theory, from which the existence of a set system that realises the structure of the natural numbers follows.

We have the impression that Hellman's system can indeed serve as a basis for arithmetic. However, we doubt that Hellman's approach can be extended to all mathematical theories that can be formulated in set-theoretical terms. In particular, modular principles corresponding to the power set axiom and choice axiom of set theory are unclear. Even more serious, in our view, is the epistemological problem of modular structuralism. When we recognise an abstract system, we thereby recognise the logical possibility of its existence. But how can the logical possibility of the existence of an abstract system be recognised without recognising the existence of the system? A proponent of modular structuralism should give us an answer to this question, and thus clarify the concept of logical possibility. Unfortunately, we find no answer to our question in the literature.

Naturalism 12

Contents

12.1 Introduction

Naturalism is an influential philosophical school of thought that leaves traces in the current media discourse: Everything that exists, including man and his work, is part of nature. If we understand nature as the opposite of the spiritual-mystical world of the supernatural, the thesis of naturalism is a principle of the Enlightenment. However, the concept of nature is understood differently in contemporary philosophy. Nature is identified with the space-time physical reality. The thesis of ontological naturalism is therefore:

Everything that exists, whether it be objects, properties, events or processes, is part of the space-time physical reality.

In particular, ontological naturalism assumes that geological, chemical, biological, sociological, economic and cultural events and processes are part of the physical reality. Furthermore, a consistent ontological naturalism directly implies that there are neither mental events and processes nor abstract objects beyond the space-time physical reality. If the concept of nature were defined more broadly, such a conclusion would not necessarily be necessary. We could reconcile ourselves to considering abstract objects, such as the natural and real numbers, and mental processes, such as thoughts and feelings, as irreducible components of nature.

© The Author(s), under exclusive license to Springer-Verlag GmbH, DE, part of Springer Nature 2025

J. Neunhäuserer, *Introduction to the Philosophy of Mathematics*, Mathematics Study Resources 22, https://doi.org/10.1007/978-3-662-72179-7_12

This position is not very widespread at present. Instead, we see in contemporary ontological naturalism a modern variant of classical materialism, which can be found, for example, in the German materialists of the nineteenth century.[1] Ontological naturalism, like classical materialism, is monistic and does not allow for metaphysical plurality.

In relation to the mind, a proponent of ontological naturalism has two options, either to deny its existence or to assert that mental processes can be reduced to suitable physical processes, and that the mind is thus naturalisable. The obvious correlations between mental and neurological processes do not, however, justify such a reduction. We doubt that thoughts and feelings are actually nothing more than neurological processes. There are robust philosophical arguments that seem to exclude an ontological reduction of mental processes to physical processes.[2] For the philosophy of mathematics, however, these arguments are not relevant. In relation to mathematics, a representative of ontological naturalism has the option of denying the existence of mathematical objects or asserting that they are concretely realised in the space-time world. We will discuss the approaches of ontological naturalism in the philosophy of mathematics in Sect. 12.4.

Methodological naturalism receives more attention than ontological naturalism in contemporary philosophy. The starting point of methodological naturalism can be formulated as follows:

The best methods we have for gaining knowledge are the methods of the sciences. These methods should be the epistemological basis of philosophy.

Which methods are recognised as scientific by naturalists is not uniform. However, all naturalists share an emphasis on empirical methodology, i.e., the justification or refutation of scientific theories through systematic observations or experiments. Other common demands of naturalism on scientific theories are generality, simplicity and elegance, as well as comprehensibility and familiarity. We see in methodological naturalism a further development of classical empiricism, which we have already discussed in Sect. 4.1. Methodological naturalism emphasises the successes of empirical research in our attempt to understand and control the world. Like classical empiricism, it rejects methods that justify theories a priori, i.e., independently of any sensory experience. Philosophical methods that rely on intuition and immediate rational insight are suspect to naturalists. The philosopher should limit himself to the application of methods that follow a scientific paradigm. Now, the principle of methodological naturalism itself is clearly normative and not descriptive. Therefore, it cannot be justified by scientific methods. Either a naturalist refrains from justifying his thesis and thus becomes a dogmatist, or he must limit his thesis and recognise the value of a philosophical theory of knowledge that justifies a limited methodological naturalism.

---

[1] Representatives of a scientific materialism in the nineteenth century were Carl Vogt (1817–1895), Ludwig Büchner (1824–1899) and also the Dutchman Jakob Moleschott (1822–1893).

[2] We have compiled some of these arguments in Neunhäuserer (2007).

In addition to epistemological problems, methodological naturalism also has significant problems with the design of a philosophy of mathematics. On the one hand, mathematical methods are constantly used in the natural sciences and also in all other empirical sciences. In contemporary science, mathematics is indispensable. On the other hand, mathematical theories, taken by themselves, are not empirical; they are not de facto confirmed or refuted by observations or experiments. We will describe in Sects. 12.3 and 12.4 how methodological naturalism attempts to make a reasonable reference to mathematics. First, we present the methodological naturalism of the influential American philosopher Willard Van Orman Quine (1908–2000) and discuss his philosophy of mathematics. Then we will discuss variants of methodological naturalism found in the American philosopher Penelope Maddy (1950–) and the American philosophers John Burgess (1948–) and Gideon Rosen (1962–).

12.2 Quine's Naturalism

In Quine's early work, we find a radical formalism and a nominalism with respect to mathematics. We have already discussed this in Chap. 9. In his later work, Quine revises his position and his philosophy of mathematics becomes part of a methodological naturalism.[3] The starting point and basis of this naturalism is a pronounced holism in epistemology, which we would like to explain first. The idea of holism is that individual scientific statements cannot be proven or refuted. It is theories, i.e., entire systems of statements, that have empirical content and are confirmed or refuted by observations and experiments. Quine speaks of the *web of our beliefs*, at the edges of which are beliefs that recur to observations and experiments. All beliefs in such a web are interconnected and none of these beliefs is unrevisable. If we revise a belief based on new empirical findings, this can in principle lead to adjustments of any other beliefs in the web. The totality of the statements of the sciences constitute for Quine a single large web, which is open for empirically motivated adjustments. Quine rejects the classical philosophical distinction between statements that are a priori true and statements that are empirically true, as well as the distinction between statements that are analytic and statements that are synthetic.[4] In the web of scientific statements, no statement is completely independent of experience and thus a priori true. Quine also holds a holism with respect to the meaning of scientific concepts. Since the meaning of scientific concepts depends on their role in scientific discourse, there are no scientific statements that are true solely on the basis of the meaning of the concepts used, i.e., analytic. Here, Quine's approach is reminiscent of structuralism, which we discussed in Chap. 11.

---

[3] Here, Quine (1981, 1990, 1995) are particularly relevant.

[4] See also Chap. 5.

For Quine, mathematics is part of the overall scientific system. Mathematical statements express beliefs that are part of the web of all our scientific beliefs. Mathematical axioms are not justified for Quine on the basis of a priori mathematical intuition or rational insight. Rather, the justification of mathematical theories and especially mathematical axiom systems lies in the justification of the overall system through observations and experiments. Thus, mathematical theories are not independent of empirical findings, but can be revised on the basis of such. Individual statements of mathematics are not a priori, i.e., true independently of any experience, since no part of the web of our scientific beliefs is completely independent of empirical findings. Ultimately, mathematical theories have empirical content by virtue of the role they play in the empirical sciences. Even logic is for Quine part of the overall scientific system and not immune to revisions. With the exception of the law of non-contradiction, Quine's logic also follows a holistic and empiricist dogma. Logical and analytical sentences, which are true solely on the basis of the meaning of the concepts used in them, have no place in Quine's methodological naturalism. In ontology, Quine starts from the principle that we should assume the existence of those objects that are indispensable in our best scientific theories. Now, some abstract mathematical objects are indeed indispensable in our best scientific theories and we should therefore assume their existence. We have already discussed this argument in Chap. 3. When Quine speaks of the sciences, he always refers to the totality of the sciences that have an empirical reference. He therefore advocates a realism with respect to the part of mathematics that finds application in the empirical sciences and is indispensable there. His attitude towards the part of mathematics that does not find application and probably will not find application remains unclear. Most of the time, he seems to ignore this part of mathematics.

Quine's naturalism is not ontological naturalism. From his ontological principle oriented towards the sciences, the existence of certain abstract objects follows just as the existence of certain concrete objects. Concrete objects, such as elementary particles, have space-time properties. Abstract objects, such as numbers, do not have these properties and are therefore not part of physical reality.

Quine's philosophy of mathematics may be original, but in our view it does not refer to mathematical practice. We even doubt whether it is a reasonable philosophical reference to mathematics at all.

It is well known that mathematicians do not make observations of a space-time nature or conduct experiments on it in their professional practice. For the researching mathematician, who defines concepts, makes conjectures and proves theorems, empirical findings appear irrelevant for their own practice. Also in the scientific community, a mathematician is generally not expected to be familiar with developments in the empirical sciences, and indeed some excellent mathematicians are not familiar with these developments. If mathematical results are revisable for empirical reasons, as Quine believes, the practice of mathematics would be fundamentally irrational in one aspect. We do not believe that Quine seriously wants to claim this.

In addition to the inadequate reference to mathematical practice, we believe we can identify a contradiction in Quine's philosophy of mathematics. His ontological principle implies the existence of mathematical objects in the Platonic sense, see again Chap. 3. Quine's concept of science entails an ontological splitting of mathematics, which we do not accept, but this is not the crucial point. The justification of the overall scientific system and thus of the mathematical theories within it is ultimately empirical for Quine. However, observations and experiments say nothing about abstract objects that are removed from the space-time physical world. In our view, Platonism in the ontology of mathematics is incompatible with empiricism in the epistemology of mathematics.

12.3 Variants of Methodological Naturalism

The methodological naturalism that we find in Penelope Maddy, John Burgess and Gideon Rosen largely agrees with Quine's naturalism. Philosophy should be guided by scientific methodology and take into account scientific results. A philosopher is a methodological naturalist if he complies with this demand, and Maddy as well as Burgess and Rosen do this.[5] However, differences in the philosophy of mathematics can be seen, which we want to highlight. The post-Quinean naturalists agree that methodological naturalism, in its scientific investigation of the sciences, should provide an explanation of the mathematical methods in the sciences and that Quine's approach is unsatisfactory in this respect.

A starting point of Maddy's philosophy of mathematics is the demand that methodological naturalism should respect mathematical practice. The entire mathematical practice, regardless of whether mathematical results find an application in the empirical sciences or not, must be taken seriously. For Maddy, the justification and criticism of the existence statements of mathematics belong to mathematical practice and should be based on internal mathematical methods, not external philosophical methods. Contemporary mathematics assumes the existence of abstract sets, relations, functions, numbers, spaces, etc., and methodological naturalism should accept these assumptions. Mathematics does not answer the ontological question of in what metaphysical sense mathematical objects exist, i.e., what their metaphysical nature is. The methodological naturalist should ignore such extra-mathematical metaphysical questions and deal with intra-mathematical justifications of existence statements. For Maddy, science as a whole is free to use all mathematical theories that are useful and effective, and does not have to deal with the abstract ontology of these theories.

Maddy recognises the purely deductive methodology of mathematical proofs and proposes a justification of logic.[6] Basic logical principles, such as the law of excluded contradiction, are obvious to us because they are fundamental to our

---

[5] See Maddy (1997) and Burgess and Rosen (1997) for this.
[6] See Maddy (2002) for this.

conceptual understanding of the empirical world. These principles are justified because we find that the simple structures they refer to are actually present in the world. Further logical principles represent idealisations. Like all scientific idealisations, they are justified by their appropriateness in a given context. In most contexts, classical logic is appropriate for Maddy; arguments for a deviating logic show that the idealisations of classical logic are inappropriate in a certain context.

Some remarks seem appropriate to us about Maddy's philosophy of mathematics. It is undoubtedly honourable that Maddy's philosophy refers to mathematical practice. We criticised Quine's lack of such a reference in Sect. 12.2. It is true that mathematical methods, namely existence proofs or disproofs, clarify in many cases which mathematical objects exist and which do not. However, fundamental existence claims in mathematics are axioms, which cannot be proven. This applies, for example, to the existence of natural numbers, the existence of the power set, or the existence of a choice function. We have already discussed this several times. We now ask ourselves whether Maddy recognises the mathematical intuition that justifies such axioms as a mathematical method. If so, she would leave behind the empirical dogma of methodological naturalism, as mathematical intuition is not an empirical method of knowledge.

We agree with Maddy that mathematics does not tell us in what ontological sense its objects exist. Unlike her, however, we believe that it is a crucial task of the philosophy of mathematics to take a stance on ontological questions. Our understanding of the philosophy of mathematics excludes a limitation to questions that can be answered with scientific methods. An agnostic attitude to ontological questions is possible, but unsatisfactory.

Finally, we would like to note that we find Maddy's conception of mathematical logic incomprehensible. The formal logic used in mathematics deals with the structures of statements or forms of statements, see Chap. 7 for this. Whether and in what context logic is applied in the world seems irrelevant for the validity of the sentences of propositional logic and predicate logic. We also cannot recognise an idealisation, as might be common in the natural sciences, in formal logic.

Let us now turn to the philosophy of mathematics of Burgess and Rosen. This agrees with Maddy's philosophy in that it takes mathematical methods and theories seriously, even if they find no applications in the empirical sciences. They explicitly criticise Quine's naturalism for favouring the empirical sciences and marginalising mathematics. For Burgess and Rosen, the totality of the sciences consists not only of the empirical sciences, but also of mathematics, and they reject any exclusion of mathematics. In the sciences, mathematics has its own method of rigorous deductive proofs and this, as Burgess and Rosen rightly point out, differs from empirical methods such as systematic observations and experiments. In the philosophy of mathematics, Burgess and Rosen accept the use of empirical methods as well as mathematical methods. This is in line with methodological naturalism, as mathematical methods are part of science. Here, a difference can be seen to Maddy's philosophy, which seems to prefer empirical methods in the scientific reference to mathematics.

Burgess and Rosen see no reason to dispense with abstract objects in the sciences. Ontological parsimony is not one of the fundamental scientific virtues for them. For the sciences, mathematical standards such as precision and consistency are more significant. The fact that abstract objects are common and useful in the sciences is sufficient for Burgess and Rosen to assume their existence. Only scientific reasons justify the rejection of the existence of an abstract object. These reasons can be contradictions that result from the assumption of the object, or an unclear definition of the object. For Burgess and Rosen, mathematics as an independent science is capable of justifying its existence claims independently of the empirical sciences. This assumption contradicts Quine's doctrine, according to which the existence of mathematical objects is justified by their indispensability in the empirical sciences. The fact that mathematical systems are used as models in the empirical sciences only shows for Burgess and Rosen that these are a convenient way to describe space-time systems.

The philosophy of mathematics of Burgess and Rosen emphasises the independence of mathematics from the empirical sciences, both in terms of its methods and its existence claims. We welcome this. However, the question arises whether Burgess and Rosen recognise mathematical methods that decide which axiom systems mathematics should use, and thus determine which mathematical objects exist. We suspect that Burgess and Rosen equally recognise all well-defined axiom systems used by mathematicians that do not lead to contradictions. This attitude is problematic, as contradictory existence claims can be derived from different axiom systems of mathematics.[7] A rationalistic theory of knowledge of mathematics is based on mathematical methods such as mathematical intuition or immediate rational insight, which justify certain axiom systems and reject other axiom systems. Supporters of methodological naturalism probably cannot accept such a theory of knowledge of mathematics, as the underlying methodology seems unscientific to them. Particularly in Anglo-Saxon philosophy, there is a deep-seated mistrust of rationalistic methodology.

12.4 Ontological Naturalism

The basic assumption of ontological naturalism in the philosophy of mathematics is:

The objects of mathematics are found in spatio-temporal nature.

This position can be characterised as non-Platonic realism with respect to mathematics; the objects of mathematics exist independently of mental processes, but are not abstract. In the nineteenth century, the influential British philosopher, politician, and economist John Stuart Mill (1806–1873) was an advocate of a naturalistic philosophy of mathematics.[8] For Mill, the laws of mathematics are natural laws, which we ultimately justify through empirical induction. Mill does not

---

[7] We have already discussed this in Sect. 3.5 on the continuum hypothesis.

[8] See Mill (1843) for more on this.

exclude the use of deductive methods in mathematics, but for him, every logical rule of inference is only as strong as its inductive justification. For Mill, the justification of the statements of arithmetic consists in the generalisation of concrete examples. The statement $2 + 1 = 3$ is thus nothing more than a generalisation of the facts that two horses and one horse together make three horses, that two people together with one person are three people, etc. The laws of geometry also refer to the concrete spatio-temporal reality for Mill. Although Mill does not believe that perfect points, lines or planes exist in nature, these objects are idealisations based on an inductive generalisation of our observations of nature. This generalisation justifies the axioms of geometry. Mill's epistemology is consistently empiricist, there is no a priori knowledge for him.

Mill explicitly rejects Platonism in the ontology of mathematics, the objects of mathematics are not abstract for him. Numbers are for Mill properties of collections of concrete objects in spatio-temporal nature. The number symbols thus refer to all collections of physical objects that have a numerical property in common. The number 2 denotes all collections of two objects in the concrete spatio-temporal reality. The objects of geometry are for Mill limit cases of real physical objects. A straight line in geometry is thus the limit case of physical objects that are almost straight lines. For Mill, the objects of geometry as limit cases are also part of spatio-temporal nature. In relation to both arithmetic and geometry, Mill is clearly an ontological naturalist.

Before we move on to a critical discussion of ontological naturalism, we would like to introduce the philosophy of mathematics of the Australian David Armstrong (1926–2014), who is considered a significant representative of ontological naturalism in the twentieth century.[9] At the centre of Armstrong's philosophy is a metaphysical materialism, coupled with a causal theory of knowledge. We gain knowledge about objects by causally interacting with them, and the objects with which we causally interact are physical.

In relation to the natural numbers, Armstrong's approach does not differ significantly from Mill's. Natural numbers are properties that are common to summaries of concrete objects. These properties are for Armstrong universals of a certain type. Real numbers are for Armstrong nothing more than ratios of concrete objects. The number π is, for example, defined as the ratio of the circumference to the diameter of every circle. Armstrong also develops a theory of classes analogous to classical set theory. All classes are concrete for Armstrong, in the sense that they are at least potentially part of spatio-temporal reality. They are thus possibilities to summarise concrete objects. Such summaries describe properties that we find in spatio-temporal nature. Armstrong also tries to concretise set-theoretic operations, like union. The composition of classes is for him a concrete operation that joins classes together. The complement of a class, i.e., the class of elements that do not belong to a class, cannot be determined in this way. The operation of negation is not concrete. Armstrong's theory of classes is incomplete in this respect and unsuitable

---

[9] See Armstrong (1997) for more on this.

as a basis for an axiomatic set theory. Armstrong does not attempt to define numbers as sets of sets, as is common in the contemporary foundation of mathematics.

At first glance, the ontology of the natural numbers of the naturalists does not lack plausibility. However, upon closer examination, this impression fades. The first problem that arises concerns the infinity of the natural numbers. To relate an ontology of mathematics only to a finite part of the natural numbers is certainly unsatisfactory. Now, there are probably only finitely many elementary particles in spatio-temporal nature. Whether there are finitely or infinitely many possible configurations of these in physical space-time is open and depends on whether space-time is infinite or continuous. If the natural numbers were indeed the summary of concrete objects and we consider all possible configurations of elementary particles in space-time as such, it follows that physical space-time is either infinite or continuous. That the ontology of mathematics decides physical problems is certainly unacceptable. The second problem of naturalism is to specify which summaries of concrete objects may be identified with natural numbers. Consider, for example, the summary of the concrete objects 1 and 2, i.e., $\{1, 2\}$. If this summary itself is identified with the number Two, $2 = \{1, 2\}$, then the number Two would be a proper part of the number Two, which seems nonsensical. Ontological naturalism can therefore generally not speak of the number of summaries of numbers. In number theory, such numbers are of great importance. This becomes clear when we think, for example, of the function $\pi(n)$, which gives the number of prime numbers less than n.

Let us now consider Armstrong's proposal of a naturalistic ontology of real numbers. Such an ontology would incidentally also imply an ontology of natural numbers, as these are embedded in the real numbers. At first glance, the identification of the number π with the ratio of circumference and diameter of circles seems reasonable. However, if Armstrong actually means concrete circles, this only yields an approximation of π. One would have to consider the ideal circles of Euclidean geometry to define π. Whether ideal circles can be identified in nature, however, is more than questionable. It also remains completely unclear how Armstrong intends to naturalise all real numbers, and thus all natural numbers, as ratios in spatio-temporal nature.

The naturalisation of geometry as an idealisation, as proposed by Mill, is hardly tenable in the light of contemporary mathematics. Mathematics today knows different geometries and which of these can serve as a model of physical space-time is not yet fully clarified. By idealising the geometry of concrete reality, one probably only obtains one of these geometries and we do not even know exactly which one. Finally, we would also like to point out that, even if the real numbers and some geometry could be naturalised, large parts of contemporary mathematics would still be excluded from naturalisation. We had already shown in Sect. 11.3 on concrete structuralism that some mathematical structures are not realisable. These structures are certainly not naturalisable. We are convinced that it is precisely the philosophy of mathematics that shows that a naturalistic metaphysics and ontology is insufficient.

Further Developments

Contents

13.1   Conceptualism and Predicativism

Conceptualism is a current position in the philosophy of mathematics, represented by the American mathematician and philosopher Nik Weaver (1969–). Even if one does not share his peculiar position, Weaver (2005) is worth reading. This position is closely related to predicativism in the foundations of mathematics, which we will first discuss.

A definition is called impredicative if it generalises over a totality to which the defined object belongs. Otherwise, a definition is called predicative. For example, *The smallest natural number that is not the sum of at most three square numbers* is an impredicative definition. We generalise over all square numbers and thus over all natural numbers, to which the defined number also belongs. In modern mathematics, there are countless impredicative definitions of mathematical objects. For example, the axiomatic definition of a set in the sense of the Zermelo-Fraenkel system is not predicative, see appendix (Chap. 14). Also, the common definitions of natural numbers by the principle of induction and of real numbers by the supremum principle are impredicative. In naive set theory, impredicative definitions, such as the set of all sets that do not contain themselves as an element, lead to contradictions, see also Chap. 7. A number of prominent mathematicians such as Henri Poincaré (1854–1912), Bertrand Russell (1872–1970), Hermann Weyl (1885–1955) and

J. Neunhäuserer, *Introduction to the Philosophy of Mathematics*,
Mathematics Study Resources 22, https://doi.org/10.1007/978-3-662-72179-7_13

Solomon Feferman (1928–2016) have therefore tried to build mathematics purely predicatively, i.e., to avoid impredicative definitions.[1] It has been shown that predicativism in mathematics brings considerable disadvantages. On the one hand, the renunciation of impredicative definitions complicates mathematical practice, on the other hand, parts of modern mathematics cannot be developed without impredicative definitions. Only really good reasons could induce us to renounce such definitions in mathematics. Paradoxes through impredicative definitions of sets are eliminated by axiomatic set theory. We are not aware of any contradictions in contemporary mathematics that arise from impredicative definitions. As Kurt Gödel has pointed out, impredicative definitions are unproblematic if we assume that the objects of mathematics exist independently of our constructions. There is nothing to suggest that totalities exist that contain objects that can only be described by referring to the totality as a whole.[2]

Only a philosopher who has the basic intuition that mathematics is our construction will be willing to accept the disadvantages of predicativism. Nik Weaver seems to be willing to do so. His conceptualism in the philosophy of mathematics has the following premise:

A construction of a mathematical object is valid if and only if it is conceptually definite, i.e., we are capable of having a complete clear mental picture of how the construction proceeds.

We assume that Weaver believes that only mathematical objects exist for which there is a valid construction. At least he explicitly rejects the assumption of the existence of sets that are axiomatically defined and not constructed. In his view, axiomatic set theory is unsuitable as a basis for mathematics, even if its axioms are intuitively plausible. To secure the existence of a set, we need a complete clear mental picture of a step-by-step construction of the set.

Weaver's concept of construction differs significantly from the concept of algorithmic construction that we had based our explanations in Chap. 10 on. A conceptually clear understanding of a mathematical construction does not imply that the construction can be performed by a Turing machine or is physically realisable in any other way. Conversely, the existence of an algorithmic construction does imply the existence of a conceptual construction in the sense of Weaver. Weaver believes that the majority of contemporary mathematics refers to objects that can be constructed in a conceptually definite way, and therefore we do not need axiomatic set theory for mathematical practice. We do not share this view. Certain objects, whose existence is secured by the power set axiom and the choice axiom, are certainly not constructible in the sense of Weaver, but they play a role in the Contemporary mathematics plays a significant role, see also Sect. 7.5. We will refrain from pursuing this matter further here and instead offer a philosophical

---

[1] See Poincaré (1906), Russell and Whitehead (1962), Weyl (1918) and Feferman (1998).

[2] See Gödel's essay in Benacerraf and Putnam (1983).

critique of the concept of construction, which underlies conceptualism. Imaginations and mental images are per se mental objects. The properties of such objects, especially their clarity and distinctness, can be known to be subjective. Whether a construction of a mathematical object is conceptually definite, i.e., the object exists, depends on the properties of mental objects according to conceptualism. Conceptualism therefore does not rule out that mathematical existence statements are subjective statements and thus represent nothing more than a person's report of their own mental mathematical processes. This consequence is unacceptable. A representative of conceptualism must therefore explain why and in what sense the clarity and distinctness of the idea of a mathematical construction is not a purely subjective mental phenomenon. We believe that this is a serious philosophical problem and Weaver gives us no information about how to solve this problem. We have encountered this problem several times. The critical rationalism, which we discuss in Sect. 13.2, proposes a solution to the problem of the subjectivity of mental constructions.

13.2 Mathematics in Critical Rationalism

Critical rationalism is an influential philosophical system of modern times, developed by the Austrian-American philosopher Karl Popper (1902–1994). Popper's basic idea is that knowledge is not fixed once and for all, but is revisable, subject to change, and should be constantly critically examined. Among many other things, critical rationalism also includes a philosophy of mathematics, which we want to present here.

For the ontology of mathematics, Popper's three-world theory is decisive.[3] The first world is the physical world, in which physical objects, events and processes are located. The second world contains the individual, subjective mental processes. The third world is the world of possible objects of thought, these are cultural objects such as theories of all kinds, myths, religions and works of art. For Popper, the objects of mathematics belong to World 3, they are therefore neither physical nor mental. So far, Popper's position agrees with the classical Platonism in the philosophy of mathematics. However, his position is only pseudo-Platonic, as he believes that World 2 produces World 3. It is our mental processes that create the cultural objects. So we produce mathematical objects, which gain an independent existence in World 3, independent of individual mental processes. When we now examine mathematical objects in World 3, our statements are not subjective, but objective. We, as subjective individuals, have collectively created an objective cultural reality. Consider, for example, the natural numbers. According to critical rationalism, we have invented or created the natural numbers, making them part of human culture. When we examine the given natural numbers, we find that they contain infinitely many prime numbers and every natural number is a product of prime numbers.

---

[3] See the presentation in Heller (2011).

These are objective statements about objects that exist in World 3. We have seen several times that mentalistic approaches in the philosophy of mathematics have a problem with the subjectivity of individual mental processes.[4] The fact that Popper introduces an independent world of cultural objects can be understood as an attempt to solve this problem.

In the epistemology of critical rationalism, the falsification and not the verification of statements is fundamental.[5] An empirical science theory that makes correct predictions is a good theory. However, it is never definitively confirmed, i.e., verified, by observations, as it is always possible to make an observation that contradicts the predictions of our theory, thus falsifying it. The falsification of a theory leads to its revision and improvement and thus to scientific progress. For critical rationalism, the falsifiability of a theory is a necessary condition for its scientific nature. Now, mathematical theories are apparently not falsifiable by empirical findings. It is only possible to falsify an axiomatised mathematical theory by a contradiction that follows from the axioms of the theory. Apart from the paradoxes that follow from a naive set theory, falsifications of mathematical theories were of minor importance for the progress of mathematics. Nonetheless, the falsification of a contemporary axiomatic system would certainly be a momentous event in the history of mathematics.

At first glance, Popper's ontology seems quite appealing to us. We enjoy the advantage of a domain of mathematics, which is neither mental nor physical, without having to make the strong assumption of the existence of abstract objects in the Platonic sense. However, if we consider the matter more closely, doubts arise as to whether the objects of mathematics are really cultural things in the broadest sense. Theories, and especially mathematical theories, are certainly part of human culture. But this does not mean that the objects with which these theories deal are part of cultural reality. Physics, for example, deals with World 1 and psychology with World 2, to speak in Popper's terminology. The fact that both the mathematical theories and the objects they deal with should have the same ontological status seems strange to us, but is probably not a real problem. Rather, it is the temporality of World 3, which makes Popper's philosophy of mathematics vulnerable. Cultural reality arises, changes and disappears with our civilisation, it is therefore temporal. This may also apply to mathematical theories, but whether it also applies to numbers and other objects of mathematics is questionable. Even though our description of the natural numbers has evolved, we do not have the impression that the natural numbers have evolved or changed over time. It is completely impossible for us to describe a temporal evolution of the natural numbers or other objects of mathematics. Although the contemporary axiomatic theory of natural numbers differs from ancient number theory in some respects, we would not know how to think about the difference in the subject area of these theories. A

---

[4] See Chap. 5 on Kant's philosophy, Chap. 8 on Brouwer's philosophy and also the last section of this chapter.

[5] See Popper (1993).

mathematical object does not seem to us to be the kind of object that is subject to change and thus temporality.

Another problem of the philosophy of mathematics of critical rationalism is that it does not allow a concept of truth for the statements of mathematical axiomatic systems. The objects that each of these systems describes are equally part of the cultural world, as long as the system has not been falsified by a contradiction. For example, there is a part of the world 3, in which every set is finite, and a part of this world, in which there are infinite sets. An axiom that demands (or denies) the existence of an infinite set cannot therefore be assigned a truth value. We have already encountered this problem in Chap. 9 on formalism.

Finally, we would like to note here that many research mathematicians have the feeling of discovering new mathematical objects and not inventing them. The correct description of an object, which a mathematician believes to have just discovered, can cause him problems. If mathematical things were indeed our creations, these phenomena would be mental illusions. One can claim this, but we shrink back from this and prefer Popper's classical Platonism to pseudo-Platonism.

13.3 Fictionalism

Fictionalism is a position in contemporary philosophy of mathematics, introduced by the American philosopher Hartry Field (1946–) and further developed by, among others, the American philosopher Mark Balaguer (1964–) and the British philosopher Mary Leng (1972–)[6] The basic assumptions of fictionalism in the philosophy of mathematics can be formulated as follows:

(1) Mathematical theories and their statements refer to abstract objects.
(2) There are no abstract objects.
(3) The statements of mathematical theories are not true.

In relation to the first thesis, a fictionalism agrees with a Platonism in the philosophy of mathematics, see Chap. 3. Mathematics thus deals with abstract objects, these are not part of the mental or physical world. In particular, the objects of mathematics are timeless. All Platonic arguments that argue against the statements of mathematics referring to concrete mental or physical objects, a functionalist in the philosophy of mathematics can adopt, see especially Chaps. 8 and 12. With the second thesis, however, fictionalism rejects the Platonic ontology of mathematical objects. It is this metaphysical assumption of Platonism that causes discomfort to many philosophers in modern times. Most philosophers believe that the third thesis follows from the first two theses of fictionalism. If the objects to which mathematics refers do not exist, the statements of mathematics are not true in the usual sense. At first glance, this seems obvious, yet some philosophers accept the first two theses of fictionalism,

---

[6] See Field (1980), Balaguer (2009) and Leng (2010).

but reject the third thesis.[7] Behind this position is a deflationary conception of truth, in which sentences can be true, even if the singular terms of the sentences do not refer to actually existing objects. We will not consider this position further, as it contradicts our basic semantic intuition and also common language usage. To claim that the sentence *The Earth is round* is true, while at the same time doubting the existence of the Earth, appears nonsensical. Similarly, it seems nonsensical to doubt the existence of the number seven and at the same time claim that the sentence *Seven is a prime number* is true.

The question now arises as to how we should understand the statements of mathematics if they do not refer to actually existing objects and are therefore not true. The starting point for this is an analogy to literary fiction. We usually assume that unicorns do not exist and the statements in theories about unicorns are not true. Nevertheless, stories about unicorns can be told, such as: There are white and black unicorns in the land of Kitemhtira. If two unicorns mate there, their children are only black if both parents are black, otherwise the children are always white. (For this reason, black unicorns rarely mate with white ones). The situation is similar in arithmetic. We find there even and odd natural numbers. If we multiply two natural numbers, the result is only odd if both numbers are odd. However, the analogy between mathematical and literary fiction should not be overestimated. Within the framework of a literary fiction, the statements *The product of two even numbers is even* and *The product of two even numbers is odd* can be equivalent, but a mathematician finds that the first statement is true and the second statement is false. A representative of fictionalism in the philosophy of mathematics, however, claims that neither of the two statements is true. He must, without using the concept of truth, describe what makes a mathematical statement a "correct" or "right" statement, and in this way distinguish the "value" of mathematical statements.

In Field's formalistic fictionalism, a mathematical proposition is fictionally correct and thus part of the mathematical narrative, precisely when it can be derived from accepted mathematical axioms using accepted rules. Fictionalism in this form does not tell us which axioms we should accept and which we should reject, after all, no axiom is true. Furthermore, for a fictionalist, there is no reason to consider only logical rules when deriving mathematical propositions. Adherents of fictionalism can use any formal rules to derive mathematical propositions from axioms. Formalistic fictionalism therefore seems to us to be a variant of formalism in the philosophy of mathematics. Propositions that make no statement about objects and have no truth value are nothing more than formulas with a certain syntactic structure. Formal rules that do not serve truth only determine a formal game with propositions. We had already dealt with formalism in the philosophy of mathematics in Chap. 9 and do not want to repeat our criticism of this position here.

The fictionalism of Balguer and also of Lang takes a different approach than that of Field. A proposition is correct in the sense of this fictionalism, precisely when it would be true under the counterfactual condition that abstract mathematical objects

---

[7] We find such a position, for example, in Azzouni (2010).

exist. For Lang, mathematics is a game in which we assume that abstract objects, in the sense of Platonism, exist. This position is disturbing. If the assumption of the existence of an object is indispensable in a mature and well-founded science, this is a good reason to assume the existence of this object. Perhaps this is even the only scientific reason that justifies the assumption of the existence of an object. In astronomy, the assumption of the existence of the planets is indispensable. This is a good reason to assume the existence of the planets. Fictionalism implicitly admits that the assumption of the existence of abstract objects in mathematics is indispensable, but still denies the existence of such objects. This position may not be contradictory, but it is bizarre. It seems to us almost as bizarre as a philosophy of astronomy that denies the existence of the planets as independent physical objects.

We would like to add an argument here that, in our view, shows that we should accept Platonism in the philosophy of mathematics if we assume that mathematics speaks about abstract objects in the Platonic sense:

(1) We should assume that the statements that we can reasonably justify are true.
(2) The statements of mathematical theories can be reasonably justified.

From (1) and (2) it follows: (3) We should assume that the statements of mathematical theories are true.

(4) A statement about an object that does not exist is not true.

From (3) and (4) it follows: We should assume that the objects of mathematical theories exist.

Only a hardened sceptic will reject the epistemological norm in the first premise of this argument. Almost all mathematicians will accept the second premise of the argument. According to our concept, a mathematician in his professional practice is actually engaged in reasonably justifying statements of mathematical theories by proving propositions. We wonder at this point whether a philosopher who seriously disputes this is capable of a reasonable reference to mathematics as a whole at all. His attitude is certainly difficult to understand. Finally, we should note that we have already discussed the premise (4) of the argument at the beginning of this section.

Assuming that mathematics refers to abstract objects, Platonism in the philosophy of mathematics is apparently much more plausible than fictionalism. We are therefore surprised to find that fictionalism is much more prominent than Platonism in the contemporary debate in the philosophy of mathematics. Perhaps this is because innovations are currently valued and expected even in philosophy. Our attitude in this respect is rather conservative.

13.4 Potentialism

The initial thesis of potentialism in the philosophy of mathematics is:

The subject area of mathematics, i.e., the mathematical universe, is never completely given, but unfolds over time through our mathematical research.

The idea of potentialism goes back to the Greek philosopher Aristotle (384–322 BC), according to whom the infinite only exists in the sense that new objects can always be added to a collection of objects. The natural numbers would thus be potentially infinite, i.e., they are not given as a whole, but we can always extend a given set of natural numbers.[8] However, contemporary set-theoretic potentialism does not refer to the natural numbers. Rather, it is about the fact that the set-theoretic universe of mathematics is never given as a whole, but unfolds piece by piece through our research. Current contributions to potentialism in the philosophy of mathematics can be found, among others, in the work of the American mathematician and philosopher Joel David Hamkins and the Norwegian philosopher Oystein Linnebo (1971–).[9] In their potentialism, they assume that the reach of the entire set-theoretic universe exists only potentially and not actually. One can always add new sets to the known set-theoretic universe as needed, but in this way one never obtains the entire set-theoretic universe. Hamkins and Linnebo distinguish three types of set-theoretic potentialism, a height potentialism, a width potentialism, and a potentialism that allows an extension of the set-theoretic universe in height and width. By this they mean that an extension of the set-theoretic universe by new ordinal numbers represents an extension of its height. Aristotle had this kind of extension in mind. The expansion of the set-theoretical universe by new subsets of infinite sets is an expansion of the universe in breadth. Behind this type of expansion is the method of so-called forcing of contemporary set theory, which allows the introduction of new cardinal numbers.[10] In any case, the mathematical universe is approximated according to potentialism by mathematical sub-universes, with the process of approximation never reaching a conclusion.

Let us now return to the initial thesis of potentialism. We can agree with it without reservation if we read it epistemologically. Mathematical research creates knowledge about the mathematical universe, which unfolds piece by piece before our mental eye. It is not to be assumed that this process will ever be completed and our mathematical knowledge is complete. The ontological position of potentialism seems less clear to us. A Platonism, which claims, that the mathematical universe consists of abstract objects and exists independently of mental processes, is compatible with our epistemological reading of potentialism. Mathematical research would be from the Platonic perspective an endless journey of discovery into an abstract

---

[8] See Aristoteles (2003).

[9] See Hamkins and Oystein (2018), Hamkins (2018) and Linnebo (2013).

[10] This matter is quite technical, we refer to Cohen (1966).

universe, which would exist even if we did not undertake this journey. Whether we may understand the set-theoretical potentialism of Hamkins and Linnebo in this way, is unfortunately anything but obvious. On the one hand, Hamkins and Linnebo talk about that the mathematical sub-universes of potentialism can be understood as independent universes, which are parts of a mathematical multiverse, This view suggests a Platonic ontology. On the other hand, for Hamkins and Linnebo, sets, which we have not yet considered in our mathematical research, exist only potentially. This is incompatible with a Platonic ontology and suggests a modal ontology. According to such an ontology, mathematical objects only exist actually when they have found their way into our mathematical theories. As long as this is not the case, the objects of mathematics exist only as possibilities. We are not sure whether Hamkins and Linnebo actually want to claim this. In any case, a modal ontology of mathematics implies that the objects of mathematics do not exist independently of mental processes, since our mathematical theories are constituted by such processes. A proponent of a modal ontology must explain to us how mental processes ensure that mathematical objects, which exist only as possibilities, become part of the actual reality. One way to do this is to assume a mathematical act of creation, as intuitionism does. Through our creations, a mathematical reality would thus be created from possibilities. We have already critically discussed this position in this section and also in Chap. 8 and also in Sect. 13.2. One might also come up with the idea that an act of cognition can serve as an explanation of the transition from the possibility of the existence of a mathematical object to its actual existence. But this is not the case. We can only recognise the possibility of the existence of a mathematical object that does not exist actually. The recognition of the possibility of the existence of an object alone is not sufficient to secure its existence. Such a recognition only shows according to modal logic that it is not necessary that the object, whose actual existence we are asking about, does not exist.

13.5 Category Theory

Category theory is not a matter of philosophy of mathematics in itself. Rather, it is a fundamental mathematical theory that has received a great deal of attention in recent decades and could potentially become relevant for the philosophy of mathematics. In textbooks, a category is usually introduced as follows:[11]

Given is a class of objects. For any two (not necessarily different) objects A, B there exists a set of morphisms $\mathbf{HOM}(A, B)$ from A to B, which are represented as arrows. $f : A \to B$ thus denotes a morphism from A to B. For three objects A, B C and morphisms $f : A \to B$ and $g : B \to C$ there exists a composite morphism $g \circ f : A \to C$. The operation $\circ$ satisfies the associative law $(f \circ g) \circ h = f \circ (g \circ h)$. Furthermore, for each object A there exists an identical morphism $id_A : A \to A$, such that $f \circ id_A = f$ and $f = id_B \circ f$ for all morphisms $f : A \to B$.

---

[11] See for example Brandenburg (2016).

The following table shows how universal categories are in mathematics:

Objects	Morphisms
Sets	Mappings
Ordered sets	Monotone mappings
Measure spaces	Measurable mappings
Topological spaces	Continuous mappings
Topological spaces	Homotopy equivalence
Differentiable manifolds	Smooth mappings
Vector spaces	Linear mappings
Algebraic structures (Groups, Rings, Fields, Algebras)	Homomorphisms
Formulas	Implications

It is undisputed that categories systematise and unify mathematics. They are an important tool for revealing connections between different mathematical disciplines and exploiting these in proofs. We find applications of category theory in algebraic topology, homological and homotopical algebra, and contemporary algebraic geometry. Now the question arises to what extent categories in mathematics are not only useful, but even fundamental. Our definition of a category refers back to the concept of sets, in the sense of this definition, sets are more fundamental than categories. However, category theorists try to dispense with the concept of sets in the definition of a category and to build set theory within category theory, thus establishing categories as fundamental objects of mathematics.[12] An extended category theory is probably equivalent to the axiomatic set theory of Zermelo and Fraenkel.[13] Nevertheless, the categorical foundation of mathematics is not undisputed. Category theorists speak in their definition of collections (*collection*) of arrows or morphisms. It is questionable whether this talk does not implicitly presuppose the concept of sets. However, we cannot conclusively answer this difficult question here.

If a categorical foundation of mathematics succeeds, it is of certain philosophical relevance. In Chap. 11 on structuralism in the philosophy of mathematics, we lacked an axiomatic definition of a structure that does not refer back to the concept of sets. If category theory provides such a definition, we would have to reassess the Platonic structuralism in Sect. 11.2 and also the eliminative structuralism in Sect. 11.4. The thesis that the fundamental objects of mathematics are structures in a categorical sense would be plausible. Perhaps the reason why a categorically structuralist foundation of mathematics seems unattractive to us is only due to our imprinting by set-theoretical formalism. Of course, this is not a good reason to reject an alternative foundation of mathematics.

---

[12] This approach goes back to Lawvere (1964).

[13] See Lane and Moerdijk (1994) and McLarty (1992) for this.

Appendix: Set Theory

14

In this appendix, we provide a brief introduction to set theory, which forms the basis of modern mathematics. Details can be found, for example, in Deiser (2010). A set, in the sense of naive set theory, is a well-defined collection of well-distinguished objects of our thought or our intuition into a whole.[1]

Sets are uniquely determined by their elements. If A is a set, then $x \in A$ means that x is an element of the set A and $x \notin A$ means that x is not in A. Two sets A and B are equal if and only if they have the same elements, i.e., x is in A if and only if $x \in B$. The empty set is denoted by $\emptyset$, it contains no elements, so $x \notin \emptyset$ for all x.

A set can be given by enumerating the elements, such as $A = \{1, 2, 3\}$ or $B = \{1, 3, 5\}$. If P is a predicate that refers to a domain U, which is a set, then

$$C = \{x \in U \mid P(x)\}$$

is a set. See Sect. 6.4 for an introduction to predicate logic. For example, consider the predicate

$$P(x) \; : \; x \text{ is a prime number,}$$

on the natural numbers $\mathbb{N}$, then

$$C = \{x \in \mathbb{N} \mid P(x)\} = \{2, 3, 5, 7, 11, 13, \dots\}$$

is the set of all prime numbers.

The union of two sets A and B is

$$A \cup B := \{x \mid x \in A \text{ or } x \in B\},$$

---

[1] This definition goes back to the founder of set theory Georg Cantor (1845–1918).

© The Author(s), under exclusive license to Springer-Verlag GmbH, DE, part of Springer Nature 2025
J. Neunhäuserer, *Introduction to the Philosophy of Mathematics*,
Mathematics Study Resources 22, https://doi.org/10.1007/978-3-662-72179-7_14

and the intersection of the sets is

$$A \cap B := \{x \mid x \in A \text{ and } x \in B\}.$$

Considering the above examples, we get $A \cup B = \{1, 2, 3, 5\}$ and $A \cap B = \{1, 3\}$ as well as $C \cap A = \{2, 3\}$ and $C \cup B = \{1, 2, 3, 5, 7, 11, 13, \dots\}$. The expressions

$$\bigcup_{i \in I} A_i, \qquad \bigcap_{i \in I} A_i$$

denote the union and the intersection of families of sets. I is here an index set, so that for each $i \in I$ a corresponding set A_i exists. For example, consider

$$G = \bigcup_{i \in \mathbb{N}} \{2i\} \text{ and } N = \bigcap_{i \in \mathbb{N}} \{2i\}$$

with the index set $\mathbb{N}$ of natural numbers, then G is the set of even numbers and N is the empty set.

A set A is a subset of a set B (in symbols $A \subseteq B$), if all x from A are also in B. The empty set is a subset of every set. The set of all subsets of B is the power set of B.

$$P(B) = \{A \mid A \subseteq B\}.$$

For example, if $B = \{1, 3, 5\}$, we get

$$P(B) = \{\emptyset, \{1\}, \{3\}, \{5\}, \{1, 3\}, \{1, 5\}, \{3, 5\}, \{1, 3, 5\}\}.$$

The concept of set in naive set theory leads to contradictions. If we assume that the set of all sets that do not contain themselves as an element, i.e., $A = \{x \mid x \notin x\}$, is a set, then x is an element of A if and only if x is not an element of A. This is the Russell's paradox, which we will discuss in Sect. 7.4. To avoid contradictions, the concept of a set is introduced axiomatically today. We present here the system of Zermelo and Fraenkel, which has become the standard in the foundations of mathematics today and consists of the following axioms:

1. **Axiom of extensionality:** Sets are equal if and only if they contain the same elements.
2. **Axiom of the empty set:** There exists a set without elements.
3. **Axiom of pair sets:** For all A and B there exists a set C that contains exactly A and B as elements.
4. **Axiom of union:** For every set A there exists a set $B =: \bigcup A$, which contains exactly the elements of the elements of A as elements. The union of two sets A and B is thus defined as $\bigcup \{A, B\}$.

5. **Axiom of infinity:** There exists an inductive set that contains the empty set and with each element x also the set $x \cup \{x\}$.
6. **Axiom of power sets:** For every set A there exists a set P, whose elements are exactly the subsets of A.
7. **Axiom of foundation:** Every non-empty set A contains an element B, such that A and B are disjoint.
8. **Axiom of separation:** If P is a predicate, then: For every set A there exists a subset B of A, which contains exactly the elements x of A for which P applies, for which $P(x)$ is therefore true.
9. **Axiom of replacement:** If A is a set and each element of A is uniquely replaced by any set, the result is a set.
   This system of axioms is often extended by the following axiom:
10. **Axiom of choice:** If A is a set of pairwise disjoint non-empty sets, then there exists a set that contains exactly one element from each element of A.

We will go into the special role of the axiom of choice in more detail in Sect. 3.5. In the philosophy of mathematics, the existence statements of axioms 5, 6 and 10 are sometimes disputed. We would like to provide the set-theoretic definition of relations and functions here, which we often assume in this book. If A and B are sets and $a \in A$ and $b \in B$, then the ordered pair (a, b) of a and b is defined as

$$(a, b) = \{\{a\}, \{a, b\}\}.$$

Pairs are defined in this way to ensure that

$$(a, b) = (c, d) \Leftrightarrow a = c \text{ and } b = d.$$

The set of all ordered pairs is the Cartesian product

$$A \times B = \{(a, b) \mid a \in A, b \in B\}.$$

Here is an example. The Cartesian product of $A = \{1, 2\}$ and $B = \{1, 2, 3\}$ is $A \times B = \{(1, 1), (1, 2), (1, 3), (2, 1), (2, 2), (2, 3)\}$.

A relation R between two sets A and B is a subset of the Cartesian product of A and B, $R \subseteq A \times B$. Here, $a \in A$ is in relation to $b \in B$, symbolically written $a R b$, if $(a, b) \in R$. For example, $R = \{(0, 1), (1, 2), (1, 3)\}$ is a relation between the sets $\{0, 1\}$ and $\{0, 1, 2, 3\}$ with $0 R 1$, $1 R 2$ and $1 R 3$. A binary relation on a set A is a subset of $A^2 = A \times A$ and an n-ary relation on a set A is a subset of the Cartesian product $A^n = A \times \cdots \times A$. For example, identity $=$ on a set A forms a relation with $R = \{(a, a) \mid a \in A\} \subseteq A^2$.

A function or mapping $f : A \to B$ is a relation between a set A and a set B, $f \subseteq A \times B$, for which the following holds:

1. For all $x \in A$ there exists a $y \in B$ with $(x, y) \in f$ (left-total)
2. $(x, y_1), (x, y_2) \in f$ implies $y_1 = y_2$ (right-unique).

For $(x, y) \in f$ one usually writes $f(x) = y$. Here is an example. Let $A = \{0, 1\}$ and $B = \{a, b\}$. $f = \{(0, a), (1, a)\}$ defines a function $f : A \rightarrow B$ with $f(0) = a$, $f(1) = a$, $f(A) = \{a\}$.

A function $f : A \rightarrow B$ is surjective or right-total if $f(A) = B$ holds. The function is injective or left-unique, if $(x_1, y), (x_2, y) \in f$ or $f(x_1) = f(x_2)$ implies the identity $x_1 = x_2$. If f is injective and surjective, then the function is bijective or reversible. If f is reversible, then the inverse function $f^{-1} : B \rightarrow A$ is given by

$$f^{-1} = \{(y, x) \in B \times A \mid (x, y) \in f\}.$$

It holds that $f^{-1}(y) = x$ if and only if $f(x) = y$. For example, the mapping $f(x) = x + 1$ on the natural numbers $\mathbb{N}$ is surjective, but not injective, since $f(\mathbb{N}) = \{2, 3, 4, 5, 6, \ldots, \}$. On the integers $\mathbb{Z}$, this mapping is bijective with $f^{-1}(x) = x - 1$. The mapping $f(x) = 1$ is neither injective nor surjective on the natural numbers. It is only on the set $\{1\}$.

References

Aczel, P., Rathjen, M.: Notes on Constructive Set Theory, Report No. 40. Royal Swedish Academy of Sciences, Stockholm (2001)

Aristoteles: Metaphysik. Akademie Verlag, Berlin (2003)

Armstrong, D.M.: A World of States of Affairs. Cambridge University Press, Cambridge (1997)

Azzouni, J.: Talking About Nothing: Numbers, Hallucinations, and Fictions. Oxford University Press, Oxford (2010)

Balaguer, M.: Platonism and Anti-platonism in Mathematics. Oxford University Press, New York (1998)

Balaguer, M.: Fictionalism, theft, and the story of mathematics. Philos. Math. **17**, 131–162 (2009)

Barrett, J.A., Byrne, P. (Hrsg.): The Everett Interpretation of Quantum Mechanics: Collected Works with Commentary. Princeton University Press, Princeton (2012)

Barrow, J.D.: New Theories of Everything the Quest for Ultimate Explanation. Oxford University Press, Oxford (2007)

Basieux, P.: Die Top Seven der mathematischen Vermutungen, rororo. Reinbek bei Hamburg (2004)

Bauer, H.: Maß- und Integrationstheorie. de Gruyter, Berlin (1998)

Benacerraf, P.: What numbers could not be. Philos. Rev. **74**, 47–73 (1965)

Benacerraf, P., Putnam, H. (eds.): Philosophy of Mathematics: Selected Readings. Cambridge University Press, Cambridge (1983)

Bishop, E.: Foundations of Constructive Analysis. Academic Press, New York (1967)

Blum, E., Blum, P., Leinkauf, T. (Hrsg.): Marsilio Ficino: Traktate zur Platonischen Philosophie. Akademie Verlag, Berlin (1993)

Bohse, H., Rosenkranz S.: Einführung in die Logik. J.B. Metzler, Stuttgart (2006)

Boolos, G.: Logic, Logic, and Logic. Harvard University Press, Cambridge (1998)

Brandenburg, M.: Einführung in die Kategorientheorie. Springer Spektrum, Berlin, Heidelberg (2016)

Brouwer, L.: Leven, kunst en mystiek. J. Waltman Jr., Delft (1905)

Brouwer, L.: De onbetrouwbaarheid der logische principes. Tijdschrift voor Wijsbegeerte **2**, 152–158 (1908)

Brouwer, L.: Über Abbildung von Mannigfaltigkeiten. Math. Ann. **71**, 97–115 (1910)

Brouwer, L.: Beweis der Invarianz der Dimensionenzahl. Math. Ann. **70**, 161–165 (1911)

Brouwer, L.: Beweis des Jordan'schen Satzes für den n-dimensionalen Raum. Math. Ann. **71**, 314–319 (1911)

Brouwer, L.: Essentially negative properties. Indagationes Math. **10**, 322–323 (1948)

Burgess, J., Rosen, G.: A Subject with No Object. Oxford University Press, Oxford (1997)

Burton, D.M.: The History of Mathematics/An Introduction. McGraw-Hill, New York (2011)

Calude, C.S., Dinneen, M.J.: Exact approximations of omega numbers. Int. J. Bifur. Chaos **17**, 1937–1954 (2007)

© The Author(s), under exclusive license to Springer-Verlag GmbH, DE, part of Springer Nature 2025
J. Neunhäuserer, *Introduction to the Philosophy of Mathematics*,
Mathematics Study Resources 22, https://doi.org/10.1007/978-3-662-72179-7

Carnap, R.: Logische Syntax der Sprache. Springer, Wien (1934)

Chiang, A.C., Wainwright, K.: Fundamental Methods of Mathematical Economics. McGraw-Hill, New York (2005)

Church, A.: An unsolvable problem of elementary number theory. Am. J. Math. **58**, 345–363 (1936)

Cohen, P.J.: Set Theory and the Continuum Hypothesis. W. A. Benjamin, New York (1966)

Cooper, S.B.: Computability Theory. Chapman and Hall, London (2004)

Curry, H.: A formalization of recursive arithmetic. Am. J. Math. **63**, 263–282 (1941)

de Spinoza, B.: Spinoza's Ethics. Princeton University Press, Princeton (2020)

Dedekind, R.: Stetigkeit und irrationale Zahlen. Vieweg, Braunschweig (1872)

Dedekind, R.: Was sind und was sollen die Zahlen? Vieweg, Braunschweig (1880)

Deiser, O.: Einführung in die Mengenlehre. Springer, Heidelberg (2010)

Descartes, R.: Abhandlung über die Methode. epubli, Berlin (2017)

Descartes, R.: Meditationes de Prima Philosophia/Meditationen über die Erste Philosophie. Reclam, Ditzingen (2018)

Diaconescu, R.: Axiom of choice and complementation. Proc. Am. Math. Soc. **51**, 176–178 (1975)

Diogenes, L.: Leben und Lehre der Philosophen (Translated from the Greek by Fritz Jürss). Reclam, Ditzingen (2009)

Eccles, J., Popper, K.: The Self and Its Brain. Routledge, London (2014)

Erdös, P.: Some remarks on set theory IV. Michigan Math. J. **2**, 169–173 (1953–1954)

Faltings, G.: The Proof of Fermat's last theorem by R. Taylor and A. Wiles. Not. AMS **42**, 743–746 (1995)

Feferman, S.: In the Light of Logic. Oxford University Press, Oxford (1998)

Field, H.: Science Without Numbers. Princeton University Press, Princton (1980)

Fischer, N. (Hrsg.): Kant und der Katholizismus. Stationen einer wechselhaften Geschichte. Herder Verlag, Freiburg (2005)

Frege, G.: Die Grundlagen der Arithmetik: eine logisch-mathematische Untersuchung über den Begriff der Zahl. Reclam, Leipzig (1884)

Frege, G.: Grundgesetze der Arithmetik. Hermann Pohle, Jena (1893)

Gabriel, G., Hermes, H., Kambartel, F., Thiel, C., Veraart, A. (eds.) Frege. Wissenschaftlicher Briefwechsel, Hamburg (1976)

Gentzen, G.: Die Widerspruchsfreiheit der reinen Zahlentheorie. Math. Ann. **112**, 493–565 (1936)

Gödel, K.: Eine Interpretation des intuitionistischen Aussagenkalküls, reproduced and translated with an introductory note by A. S. Troelstra in Gödel 1986, 296–304 (1933)

Gödel, K.: The Consistency of the Continuum-Hypothesis. Princeton University Press, Princeton (1940)

Goodman, N.D., Myhill, J.: Choice implies excluded middle. Zeitschrift für Mathematische Logik und Grundlagen der Mathematik **24**, 461 (1978)

Goodman, N., Quine, W.V.: Steps towards a Constructive Nominalism. J. Symb. Logic **12**, 97–122 (1947)

Hale, B., Crispin, W.: The Reasons Proper Study: Essays Towards a Neo-Fregean Philosophy of Mathematics. Oxford University Press, New York (2001)

Hamkins, J.D.: The set-theoretic multiverse. Rev. Symb. Log. **5**(3), 416–449 (2012)

Hamkins, J.D.: The modal logic of arithmetic potentialism and the universal algorithm. arxiv:1801.04599 (2018)

Hamkins, J.D., Oystein L.: The modal logic of set-theoretic potentialism and the potentialist maximality principles. arXiv:1708.01644 (2018)

Heck, R.: Reading Freges Grundgesetze. Clarendon Press, Oxford (2012)

Hegel, G.W.F.: Werkausgabe. Suhrkamp, Frankfurt am Main (1970)

Heitsch, E. (Hrsg.): Xenophanes: Die Fragmente. Akademie Verlag, Berlin (2014)

Heller, M.: Philosophy in Science: An Historical Introduction. Springer, New York (2011)

Hellman, G.: Mathematics Without Numbers. Oxford University Press, Oxford (1989)

Heuser, H.: Gewöhnliche Differentialgleichungen: Einführung in Lehre und Gebrauch. Vieweg+Teubner Verlag, Wiesbaden (2009)

Heyting, A.: Die formalen Regeln der intuitionistischen Logik, Sitzungsberichte der Preussischen Akademie der Wissenschaften. Physikalisch-mathematische Klasse, 42–56 (1930)

Hilbert, D.: Neubegründung der Mathematik: Erste Mitteilung. Abhandlungen aus dem Seminar der Hamburgischen Universität **1**, 157–177 (1922)

Hilbert, D.: Die Grundlagen der Mathematik. Abhandlungen aus dem Seminar der Hamburgischen Universität **6**, 65–85 (1928)

Hilbert, D.: Die Grundlegung der elementaren Zahlenlehre. Math. Ann. **104**, 485–494 (1931)

Hilbert, D.: Gesammelte Abhandlungen, vol. 3. Springer, Berlin (1935)

Hilbert, D., Bernays, P.: Grundlagen der Mathematik 1+2. Springer, Berlin (1934/1939)

Hoffmann, P.: Der Mann, der die Zahlen liebte. Ullstein, Berlin (1999)

Hoffmann, D.W.: Theoretische Informatik. Carl Hanser Fachbuchverlag, München (2022)

Hübener, W.: Ockham's Razor not Mysterious. Archiv Begriffsgeschichte **27**, 73–92 (1983)

Hume, D.: A Treatise of Human Nature. Oxford University Press, Oxford (2000)

Johannson, I.: Der Minimalkalkül, ein reduzierter intuitionistischer Formalismus. Compos. Math. Band **4**, 119–136 (1936)

Kant: Ausgabe der Preußischen Akademie der Wissenschaften (1900–1908), Nachdr. Walter de Gruyter, Berlin (1962)

Kaye, R.: Models of Peano Arithmetic. Oxford University Press, Oxford (1991)

Keil, G., Schnädelbach, H. (Hrsg.): Naturalismus. Philosophische Beiträge. Suhrkamp, Frankfurt (2000)

Kleene, S.C.: Introduction to Metamathematics. North-Holland Publishing Company, Amsterdam (1991)

Kolman, E., Yanovskaya, S.: Hegel & Mathematics. In: Yanovskaya, S. (ed.) (1983) Marx's Mathematical Manuscripts. New Park Publications, London (1931)

Kuhn, W.: Ideengeschichte der Physik: Eine Analyse der Entwicklung der Physik im historischen Kontext. Springer Sektrum, Heidelberg (2016)

Kunen, K.: Set Theory: An Introduction to Independence Proofs. Elsevier, Amsterdam/North-Holland (1980)

Lane, S.M., Moerdijk, I.: Sheaves in Geometry and Logic: A First Introduction to Topos Theory. Springer, New York (1994)

Lawvere, W.: An elementary theory of the category of sets. Proc. Natl. Acad. Sci. U.S.A. **52**, 1506–1511 (1964)

Leibniz, G.W.: Monadologie. Contumax, Berlin (2017)

Leibniz, G.W.: Sämtliche Schriften und Briefe. De Gruyter Akademie, Berlin (2019)

Leng, M.: Mathematics and Reality. Oxford University Press, Oxford (2010)

Linnebo, O.: Structuralism and the Notion of Dependence. Philos. Q. **58**, 59–79 (2008)

Linnebo, O.: The Potential Hierarchy of Sets. Rev. Symb. Logic **6**(2), 205–228 (2013)

Loewenthal, E. (Hrs.) Platon (Autor): Sämtliche Werke in drei Bänden. Lambert Schneider Verlag, Darmstadt (2014)

Ludwig, G.: Einführung in die Grundlagen der theoretischen Physik. Vieweg Verlag, Wiesbaden (1985)

Maddy, P.: Naturalism in Mathematics. Clarendon Press, Oxford (1997)

Maddy, P.: A naturalistic look at logic. Proc. Addresses APA **76**, 61–90 (2002)

Mandelbrot, B.B.: Die fraktale Geometrie der Natur. Springer, Basel (2014)

Mansfeld, J.: Die Vorsokratiker I. Reclam, Ditzingen (1986)

Markov, A.A.: Theory of Algorithms. Trudy Mat. Istituta imeni V. A. Steklova, Lenningrad (1954)

Martin-Löf, P.: An intuitionistic theory of types: predicative part. In: Rose, H.E., Shepherdson, J.C. (eds.) Logic Colloquium. North-Holland, Amsterdam (1973)

Matijassewitsch, Y.: Hilberts 10th Problem. The MIT Press, Cambridge (1993)

McLarty, C.: Elementary Categories, Elementary Toposes. Oxford University Press, Oxford (1992)

Mill, J.S.: A System of Logic. Forgotten Books, London (1843)

Mines, R., Richman, F., Ruitenburg, W.: A Course in Constructive Algebra. Springer, Heidelberg (1988)

Mohr, P.J., Taylor, B.N., Newell, D.B.: CODATA recommended values of the fundamental physical constants. Rev. Mod. Phys. **80**(2), 633–730 (2008)

Mollweide, K.B., Lorenz, J.F. (eds.) Euklid. ChiZine Publications, Toronto (2017)

Monk, D.: Mathematical Logic. Springer, Berlin (1976)

Nagel, E., Newman, J.R.: Der Gödelsche Beweis. Scientia Nova, Oldenburg (2003)

Neunhäuserer, J.: Wider die Materialistische Metaphysik, Marburger Forum, Jg. 8, Heft 4, Marburg (2007)

Neunhäuserer, J.: Schöne Sätze der Mathematik. Springer Spektrum, Berlin (2015)

Neunhäuserer, J.: Mathematische Begriffe in Beispielen und Bilder, Springer Spektrum, Berlin, Heidelberg (2017)

Newton, I.: Mathematical Principles of Natural Philosophy. University of California, Berkeley (2016)

Parsons, C.: Freges theory of number. In: Black, M. (ed.) Philosophy in America. Cornell University Press, New York (1965)

Parsons, C.: Platonism and mathematical intuition in Kurt Gödel's thought. Bull. Symb. Logic **1**, 44–74 (1995)

Peano, G.: Arithmetices principia, nova methodo exposita. Fratres Bocca, Turin (1889)

Penrose, R.: The Emperor's New Mind. Oxford University Press, Oxford (2016)

Picado, J., Pultr, A.: Frames and Locales: Topology without Points. Birkhäuser Verlag, Basel (2011)

Pinkard, T.: Hegel's Philosophy of mathematics. Philos. Phenomenol. Res. **41**(4), 452–464 (1981)

Poincaré, H.: Les Mathématiques et la Logique. Rev. Metaphys. Morale **14**, 294–317 (1906)

Pollok, K.: Die Vereinten Nationen im Lichte Immanuel Kants Schrift Zum ewigen Frieden, Sic et Non (1996)

Popper, K.R.: What is dialectic. Mind **49**(196), 403–426 (1940)

Popper, K.: Objektive Erkenntnis. Hoffmann und Campe, Hamburg (1993)

Putnam, H.: Philosophy of Logic. Harper Torch Books, New York (1971)

Quine, W.V.O.: Set Theory and Its Logic. Harvard University Press, Cambridge (1963)

Quine, W.V.O.: Theories and Things. Harvard University Press, Cambridge (1981)

Quine, W.V.O.: Pursuit of Truth. Harvard University Press, Cambridge (1990)

Quine, W.V.O.: From Stimulus to Science. Harvard University, Cambridge (1995)

Rautenberg, W.: Einführung in die mathematische Logik. Vieweg+Teubner Verlag, Wiesbaden (2008)

Resnik, M.: Mathematics as a Science of Patterns. Oxford University Press, Oxford (1997)

Rosenthal-Schneider, I.: Begegnungen mit Einstein, von Laue und Planck. Vieweg Verlag, Braunschweig/Wiesbaden (1988)

Russel, B.: History of Western Philosophy. Routledge, London (2004)

Russell, B., Whitehead, A.N.: Principia Mathematica. Cambridge University Press, Cambridge (1962)

Schechter, E.: Handbook of Analysis and Its Foundations. Academic Press Inc., London (1997)

Schelling, F.W.: Cotta, Stuttgart (1856–1861)

Schlegel, F.: Kritische Ausgabe seiner Werke, Paderborn (1958ff)

Shapiro, S.: Philosophy of Mathematics: Structure and Ontology. Oxford University Press, New York (1997)

Shapiro, S. (Hrsg.): The Oxford Handbook of Philosophy of Mathematics and Logic. Oxford University Press, Oxford (2007)

Sierpinski, W.: Cardinal and Ordinal Numbers. Polish Scientific Publishers, Warschau (1965)

Specker, E.: Nicht konstruktiv beweisbare Sätze der Analysis. J. Symb. Logic **14**, 145–158 (1949)

Stegmüller, W. (Hrsg.): Das Universalien-Problem. WBG, Darmstadt (1978)

Störig, H.J.: Kleine Weltgeschichte der Philosophie. Kohlhammer Verlag, Stuttgart (2016)

Tennant, N.: Natural logicism via the logic of orderly pairing. In: Logicism, Intuitionism, Formalism: What Has Become of Them? Synthese Library, Springer (2009)

Troelstra, A.S., van Dalen, D.: Constructivism I and II. North-Holland, Amsterdam (1988)

Turing, A.: On computable numbers, with an application to the Entscheidungsproblem. Proc. Lond. Math. Soc. Ser. 2, **42**, 230–265 (1936)

van Atten, M.: Brouwer Meets Husserl (On the Phenomenology of Choice Sequences). Springer, Dordrecht (2007)

van Dahlen, D.: Mystic, Geometer, and Intuitionist: The Life of L. E. J. Brouwer. Clarendon Press, Oxford (1999)

Weaver, N.: Mathematical Conceptualism. arXiv:math/0509246 (2005)

Weyl, H.: Das Kontinuum. Verlag von Veit und Comp., Leipzig (1918)

Wigner, E.: The unreasonable effectiveness of mathematics in the natural sciences. Commun. Pure Appl. Math. **13**, 1–14 (1960)

Wittgenstein, L.: Philosophical Grammar. Blackwell, Oxford (1975)

Wittgenstein, L.: Tractatus logico-philosophicus, Logisch-philosophische Abhandlung. Suhrkamp, Frankfurt am Main (2003)

Woddin, W.H.: The continuum hypothesis I/II. Not. AMS Bd. **48**(6/7) (2001)

Wright, C.: Freges Conception of Numbers as Objects. Aberdeen University Press, Aberdeen (1983)

Zalta, E.: Natural numbers and natural cardinals as abstract objects: a partial reconstruction of Freges Grundgesetze in object theory. J. Philos. Logic **28**(6), 619–660 (1999)

Zalta, E.N. (Hrsg.): The Stanford Encyclopedia of Philosophy (2017 edition). Stanford University, Stanford (2017). https://plato.stanford.edu/

Person Index

K

Kant (1724–1804), 30, 39
Kepler (1571–1630), 10
Kolmogorov (1903–1987), 74
Kronecker (1823–1891), 26

L

Leibniz (1646–1716), 30, 36, 40, 57
Linnebo (1971–), 115, 136
Locke (1632–1704), 30

M

Maddy (1950–), 121, 123
Markov junior (1903–1979), 100
Martin-Löf (1942–), 101
Mill (1806–1873), 125

N

Newton (1642–1726), 30, 34

O

Ockham (1288–1347), 26

P

Parsons (1933–), 59
Peano (1858–1932), 57
Penrose (1931–), 32
Plato (428–348 BC), 13, 30
Plotinus (205–270 BC), 13
Poincaré (1854–1912), 130
Polykrates (about 570–522 BC), 5
Popper (1902–1994), 131
Putnam (1926–2016), 21
Pythagoras (about 570–495 BC), 5

Q

Quine (1908–2000), 21, 90, 121

R

Reichenbach (1891–1953), 89
Resnik (1938–), 111
Rosen (1962–), 121, 123
Russell (1872–1970), 57, 130

S

Schelling (1775–1854), 14, 41, 48
Schlegel (1772–1829), 52
Schlick (1882–1936), 89
Shapiro (1951–), 111
Socrates (469–399 BC), 13
Spinoza (1632–1677), 30

T

Taylor (1962–), 105
Tennant (1950–), 59
Thales (about 624–546 BC), 5
Thomae (1840–1921), 89
Turing (1912–1954), 92, 100

V

van Dalen (1932–), 74
van Schooten (1615–1660), 32
von Neumann (1903–1957), 111

W

Weaver (1969–), 129
Weyl (1885–1955), 130
Whitehead (1861–1947), 57
Wigner (1902–1995), 11
Wiles (1953–), 105
Wittgenstein (1889–1951), 89
Woodin (1955–), 26
Wright (1942–), 59

Z

Zalta (1952–), 59
Zermelo (1871–1953), 16, 58, 111

Subject Index

A

Abacus, 91
Absolute ego, 47
Absolute identity, 48
Absolute space, 35
Abstraction, 112
Abstract object, 15
Abstract structuralism, 111
Abstract structure, 112
Acausal, 17
Act of creation, 137
Actually infinite, 79
Algebraic geometry, 34
Algebraic number, 99
Algebraic structure, 110
Algorithm, 86, 98
Analysis, 34, 36
Analytically a priori, 56
Analytic geometry, 32
Analytic judgements, 40
Ante-rem structuralism, 111
Anti-zero, 68
Aperion, 8
Apollonian problem, 32
A posteriori, 40
A priori, 30, 35, 37, 40
A priori knowledge, 30
Aristotelian structuralism, 114
Arithmetic, 42, 56
Axiom, 56
Axiomatic set theory, 20, 23, 64
Axiom of choice, 69, 101, 141
Axiom of infinity, 65
Axiom schema, 63
Axioms of mathematics, 23

B

Basic intuitions, 24
Basic Law V, 64
Being-in-and-for-itself, 50
Being-in-itself, 50
Being-other, 50
Bertrand's postulate, 88
Biological evolution, 11
Brain research, 32
Brouwer-Heyting-Kolmogorov Interpretation, 77
Brouwer's fixed point theorem, 74

C

Calculus, 34, 86
Cardinality, 68
Cardinal numbers, 68, 102
Cartesian coordinate system, 32
Cartesian doubt, 31
Cartesian product, 141
Category, 138
Category theory, 137
Cauchy sequences, 83
Causal relations, 23
Causal theory of knowledge, 126
Causal theory of truth, 23
Chaitin constant, 99
Characteristica universalis, 37
Choice sequences, 80
Church-Turing thesis, 98
Class, 19, 127
Classical logic, 59
Classical mechanics, 34
Complete induction, 63, 80

FSC
www.fsc.org
MIX
Papier aus verantwortungsvollen Quellen
Paper from responsible sources
FSC® C105338